MIXING FOR THE PROCESS INDUSTRIES

MIXING FOR THE PROCESS INDUSTRIES

Robert J. McDonough

VNR VAN NOSTRAND REINHOLD
New York

Library of Congress Catalog Card Number 91-26746
ISBN 0-442-23434-1

Manufactured in the United States of America

Published by Van Nostrand Reinhold
115 Fifth Avenue
New York, New York 10003

Chapman and Hall
2-6 Boundary Row
London, SE1 8HN, England

Thomas Nelson Australia
102 Dodds Street
South Melbourne 3205
Victoria, Australia

Nelson Canada
1120 Birchmount Road
Scarborough, Ontario, M1K 5G4, Canada

16 15 14 13 12 11 10 9 8 7 6 5 4 3 2 1

Library of Congress Cataloging-in-Publication Data

McDonough, Robert J.
Mixing for the process industries / Robert J. McDonough.
p. cm.
Includes bibliographical references (p.) and index.
ISBN 0-442-23434-1
1. Mixing machinery. I. Title.
TP156.M5M38 1991
660'.284292—dc20 91-26746
CIP

To
My Wife, Kathleen and children,
Kevin and Jennifer, for their patience,
sacrifice and encouragement.

Contents

Preface

This book is written primarily for an audience of engineers who design, specify, procure, update, and maintain mixers in the processing industries. It is an application-oriented handbook to support the design and operation of mixers for liquids, liquids and solids, liquids and gases, and any combination thereof. This book does not address solids mixing.

Mixing applications are identified as either flow controlling or shear rate controlling, and are addressed as such. Specific mixing applications and design parameters are presented for blending, solid suspension, heat transfer, mass transfer, dispersions, extractions, etc.

Fundamentals of mixer design and operation, agitation applications spectrum, types and geometries of mixing impellers, etc. are presented in Chapter 1, providing the basis for practical information presented in subsequent chapters.

Flow controlled mixing applications are discussed in Chapter 2. These include blending, solid suspension, heat transfer, dissolving of solids, and crystallization. Optimization of mixer design is emphasized, with an eye to saving power and capital cost. Also discussed is the use of draft tube circulator mixers to optimize mixer performance and cost.

Shear controlled mixing applications are addressed in Chapter 3. These include gas-liquid dispersions, mass transfer, fermentation, emulsions, extractions, and dispersions of solids in liquids.

Chapter 4 focuses on geometric and nongeometric scaleup of mixers to accomplish equal or better mixing results in various scales of volume. Scaleup relevant to applications of blending, solids suspension, and mass and heat transfer is explained.

Chapter 5 is devoted entirely to static pipeline mixing. A step-by-step approach to design of static mixers is offered.

Retrofitting of agitators is presented in Chapter 6 as a cost-effective alternative to purchasing and installing a new mixer. Retrofitting of impellers is discussed from the perspective of improving mixing results at similar power levels and obtaining similar mixing results at reduced power levels.

These early chapters emphasize process-results aspects of mixer design and operation. Chapter 7 introduces mechanical design considerations for agitator design, operation and maintenance. Mixer shaft, impeller, and drive mechanical design parameters and contraints are explored. Chapter 7 also surveys the various types of commercially available mixer shaft seals and drive assemblies.

Formulae and their practical use in example calculations are found in every chapter to help the reader work through a design scenario or solve a mixing problem.

Mixing for the Process Industries will be a tool to design new mixers or revise existing agitators for optimum performance in a given application. It will enable the reader to solve operational problems, enhance mixing process results, and ensure mechanical reliability and integrity of agitators. Finally, it will provide a means to optimize mixer design and selection from an applications perspective, tried and proven in industrial processing situations.

1

Fundamentals of Mixing

A. INTRODUCTION

In the design and operation of a full-scale plant, agitation equipment often plays a major role in optimization of the overall process. To meet the overall process objectives, it is essential to determine the significant mixing parameter(s) and incorporate that knowledge into a successful commercial-scale design. Economic benefits, such as minimum power usage or low capital cost, can be achieved by applying technology in a manner that assures the specific process in mind is optimized with respect to agitation, and that the ultimate mixer selection is an optimum design.

The effect that agitation has on process results is established by evaluating the following:

1. Process Design Requirements
2. Mixer Impeller Flow and Shear Development
3. Agitator Mechanical Design Considerations
4. Vessel and Mixer Support Structural Requirements

Power consumption by a mixer with a particular impeller geometry, in a specific process and vessel environment, bears no relationship to achieving a desired process result. Flow and shear generated by mixer impellers, within a specific vessel constraint, achieve desired results for the process or application at hand. The proper balance of flow and shear development, in concert with a mixer turbine design and selection that provides this proper balance with minimum power requirement, assures an optimum agitator design. As one might expect, there is a wide spectrum of flow and shear characteristics available from a number of possible impeller or turbine configurations that must be tailored to a specific process result and mixing requirement.

Well established mixing technology can be applied to agitator design and selection for many processes such as blending, heat transfer and solids suspension. An expansion of this basic technology is required for applications involving specific mass transfer needs and many complex reactor systems. In some cases, pilot plant testwork must be pursued to develop data for process optimization, mixer scaleup criteria, and optimum mixer design and selection.

Fluid mixing can be a simple process, with every aspect of mixer performance easily evaluated. It can also be extremely complex, with many independent steps involving a variety of fluid phenomena.

Mixer performance is typically expressed in terms of bulk fluid velocity generated, shear rate, total pumping capacity of the impeller, total flow in the vessel, or in terms of a desired process mixing result (i.e., blend time, solids suspension regime, etc.).

If the application is relatively simple to evaluate, a thorough examination of complicated fluid mechanics in a mixing tank will not be necessary. If the process is complex, however, a complete investigation of fluid shear rates and stresses, flow rate, turbulence, and their effects on mixing results may be required. Such a procedure is best carried out in a controlled pilot plant test program.

It may be possible to express a complex mixing process in terms of only one, or a few, simple velocity and pumping capacity relationships. If such were the controlling factors in the process, the design of an appropriate mixing system would be straightforward.

Problems can often arise in clearly differentiating between processes to which comparatively simple physical and visual concepts can be applied and those for which more elaborate qualitative and quantitative aspects must be explored.

For example, consider a mixing process that requires the suspension of solids in a liquid, in which a chemical reaction takes place between components of the liquid and solid phases. If the process engineer decides that the reaction kinetics occur when a certain degree of solids suspension has been achieved, the engineer should issue process specifications calling only for a specific degree of solids suspension.

If the mixing system so specified accomplishes the desired process result, the agitator design will be acceptable. The opposite will be seen, however, if the observed relationship between the specified suspension of solids in the tank and completion of reaction is not as anticipated.

B. FLUID MECHANICS—FLOW AND SHEAR

Impeller type mixers are essentially pumps (albeit not very efficient ones), having many of the same characteristics.

The most fundamental of these is that the mixer impeller power consumption is proportional to its flow and head development:

$$P \propto QH \tag{1}$$

where: P = power
Q = flow rate
H = head.

All the power supplied from the agitator to the fluid produces flow, velocity head, or shear.

Unlike a pump casing, however, a mixing tank is not a confining channel. Therefore, the measurement and calculation of turbine flow and head is not as quantitative as is the case of a pump and piping system. Axial flow mixer impellers operating in a draft tube at close clearance to the tube walls simulate an axial flow pump in its casing. In this system, flow and head development are quantitative, and the efficiency in producing same per expended unit of power consumption is increased over that in a mixing tank without draft tube. Usually, impeller flow is expressed as the pumping capacity normal to the discharge plane of a turbine.

The discharge area of a typical constant pitch axial flow turbine is a conical surface, because the flow has both axial and radial components. The flow from streamlined airfoil impellers of variable blade pitch tends to be more completely axial, so its discharge plane approximates a horizontal disk surface. An impeller in a tank emits a jet stream. This stream has a certain velocity, flow and cross-sectional area. As the jet expands within the confines of the tank, the area, flow, and velocity change markedly and additional fluid is entrained. Fluid momentum, however, is conserved. The momentum of the stream leaving the impeller equals the momentum of fluid in the expanded jet. The momentum of the impeller, which manifests itself as velocity of motion in the tank, is a measurement of mixing or process result.

Impeller flow refers only to the flow directly produced by the impeller. There is also entrained flow, which is that fluid set into motion by the turbulence of the impeller stream. Entrained flow is often a major portion of the total flow, as illustrated in Fig. 1-1.

Equation (1) may be expressed in terms of impeller rotational speed N and impeller diameter D as follows:

$$P \propto N^3D^5 \quad \text{for turbulent flow regimes} \tag{2}$$

$$P \propto N^2D^3 \quad \text{for laminer flow regimes} \tag{3}$$

Other important relationships are:

$$Q \propto ND^3 \tag{4}$$

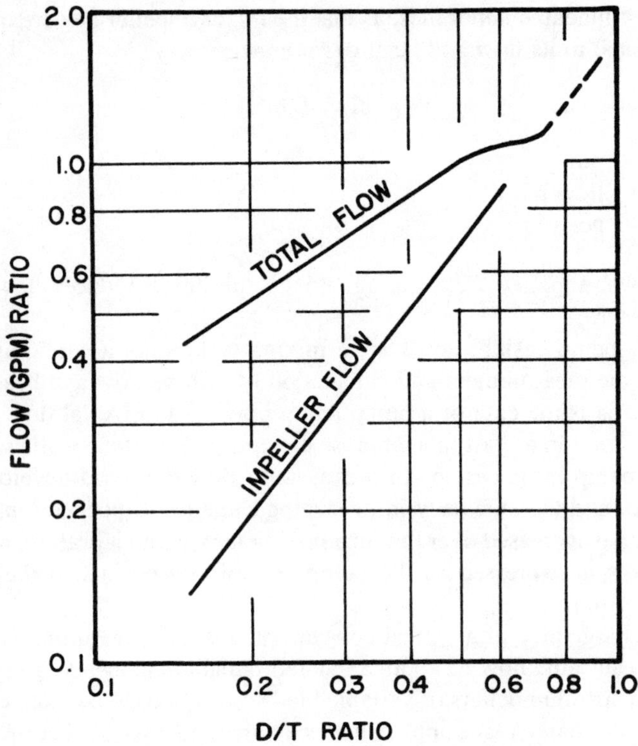

FIGURE 1-1. Total flow versus impeller flow. (*Mixing Equipment, Inc.*)

$$H \propto (ND)^2 \tag{5}$$

$$Q_p \propto D^{4/3} \tag{6}$$

$$N_p \propto D^{-5/3} \tag{7}$$

Equations (6) and (7) are derived from Eq. (1) and express Q and N at constant power (denoted by subscript p) in terms of D. As impeller diameter increases, the impeller flow increases and the speed decreases rapidly at constant power. It follows from Eq. (6) that mixing or process results predicated on flow (i.e., flow controlled applications) can be improved at constant power by using a larger diameter impeller running at a slower speed.

Another consideration is the affect that a larger diameter impeller operating at reduced speed and at constant power has on torque. Torque is related to power P and speed N as follows:

$$\text{Torque} \propto P/N \tag{8}$$

As diameter increases and speed decreases at constant power to obtain greater flow-controlled process results, the torque rises rapidly:

$$\text{Torque}_p \propto D^{5/3} \tag{9}$$

As shown in Fig. 1-2, there are numerous combinations of power and torque accomplishing the same mixing results for flow controlled applications. The question remains as to which selection is optimum for you?

The answer lies in economics. In buying an agitator, both capital cost (i.e., purchase price of the machine) and operating cost (i.e., power requirement) must be taken into account. It must be remembered that

$$\text{Capital Cost} \propto \text{Torque} \tag{10}$$

$$\text{Operating Cost} \propto \text{Power} \tag{11}$$

Your economic situation will dictate which selection is right for you. Inevitably the tradeoff between higher capital cost for lower operating cost or vice versa must be evaluated to select the optimum agitator.

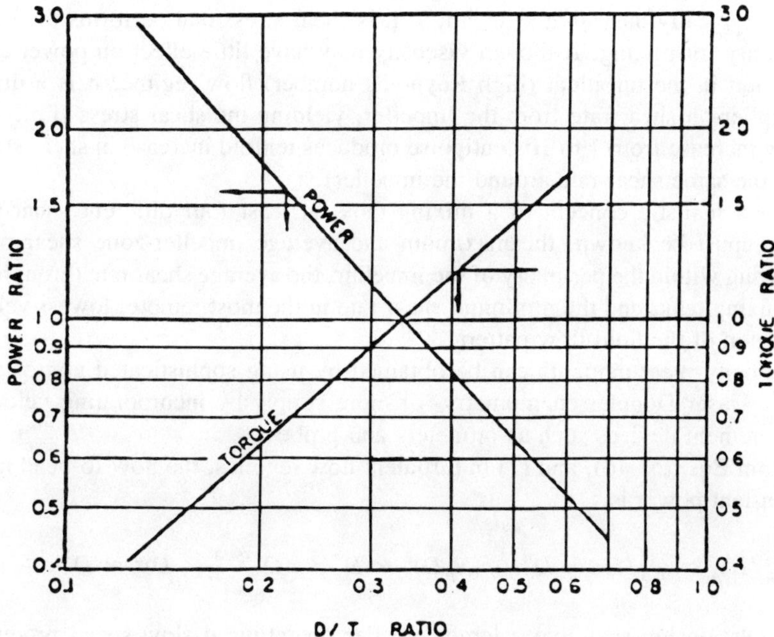

FIGURE 1-2. Power–torque relationship for constant process results at constant power input. (*Mixing Equipment, Inc.*)

A rotating mixer impeller will produce a flow velocity profile along the length of impeller blade. The profile may be nearly uniform or variable from point to point along the blade. The profile consists of velocity vectors with magnitude and direction. The direction will be predominantly axial or radial (i.e., parallel or perpendicular to the axis of rotation) depending on the impeller geometry. There is a center of rotation about which all the flow can be considered to pass.

When the flow is turbulent, velocity fluctuations are significant. These can be analyzed mathematically to give one value of average velocity at a given point in a stirred tank. When the flow is laminar, fluctuating velocity components are not present. In qualitative terms, the shear rate between adjacent layers of fluid, based on average velocity at any profile point, is used to develop the concept of macroscale shear rate. The shear operates on particles that are too large to respond to the high-speed velocity fluctuations.

Using either average velocity in turbulent flow or actual velocity in laminar flow, a velocity profile may be established for a particular impeller geometry of given diameter and speed. Measuring the slope of the velocity gradient along the blade length gives the shear rate, in reciprocal seconds, at any point on the profile. The maximum shear rate and average shear rate around the impeller can be calculated.

The product of shear rate at a given point and viscosity gives fluid shear stress (typically measured in lb/in.^2). It is shear stress that performs the work necessary for mixing. Although viscosity may have little effect on power consumption in the turbulent (high Reynolds number) flow regime, it is a direct multiplier on shear rate from the impeller, yielding the shear stress (i.e., viscosity increase from 1 to 10 centipoise produces tenfold increase in shear stress from the same shear rate around the impeller).

For a realistic concept of a mixing tank, at least four different shear-rate values must be known: the maximum and average impeller-zone shear rates occurring within the periphery of the impeller, the average shear rate throughout the mixing tank, and the minimum shear rate in the most remote, lowest-velocity region of the fluid flow pattern.

Velocity measurements can be obtained by using sophisticated approaches such as laser Doppler anemometry, or more simply by incorporating velocity measurement devices such as ottmeters and probes.

From Eqs. (5), (6), and (7) in turbulent flow regimes, the flow-to-head ratio at constant power is

$$(Q/H)_p \propto Q_p/H_p \propto D^{4/3}/(N_p D)^2 \propto D^{4/3}/(D^{-5/3} \times D)^2 \propto D^{8/3} \qquad (12)$$

This relationship says that a large impeller operating at slow speed produces high flow and low impeller head. Impeller head, being related to the square root of fluid shear rate, is a measure of the flow-to-fluid shear rate around the impeller. Hence it follows that a small impeller turning at high speed develops a high shear rate and low pumping capacity. Every mixing application has an optimum balance of flow and fluid shear.

For a given impeller geometry, *maximum* shear rate is proportional to impeller tip speed:

$$\textit{Maximum}\ \text{Shear Rate} \propto \pi DN \tag{13}$$

whereas *average* shear rate is proportional to speed:

$$\textit{Average}\ \text{Shear Rate} \propto N \tag{14}$$

Fluid discharge from an impeller can be measured with a device that has a high-frequency response, allowing velocity to be determined as a function of time. At any point in time, the fluid velocity can be expressed as an average velocity plus some fluctuating velocity component. Integration of the average velocities across the discharge of the impeller allows for the calculation of the impeller or primary pumping capacity normal to the discharge plane. This plane is bounded by the impeller blade diameter and height.

Velocity gradients between the average velocities operate only on larger particles (typically greater than 1,000 micron in size). This phenomenon is called *macroscale mixing*.

The fluctuating velocity gradients have an effect on smaller particles. In the turbulent mixing regime, these fluctuations are attributed to a finite number of impeller blades passing a finite number of tank baffles. For particle sizes typically less than 100 microns in size, turbulent properties of the fluid become an important consideration. Microscale mixing occurs in this regime.

As previously noted, all the power applied by an agitator to the fluid through the mixing impeller produces flow, velocity head, or shear. Through viscous shear, power is converted to heat at the rate of approximately 2,500 Btu/hour per Hp. Viscous shear, created only in the turbulent flow regime, is present at the microscale mixing level. Hence mixer horsepower per unit volume is the dominant component in microscale agitation. At very small particle sizes (less than, say, 1 micron), impeller geometry and type is of no consequence to viscous shear development and its effect on the particles—only the application of mixer power to generate shear is important. Experiments have shown that the power per unit volume in the impeller zone is about 100 times greater than in the rest of the tank.

Mixers are not specified for industrial applications to meet fluid mechanics parameters. Processes are so complex that it is impossible to isolate and define the effect of fluid mechanics on process results and mixer requirements. Pilot plant testing and experimentation affords a study of the sensitivity of a process to macroscale mixing variables (as a function of power, pumping capacity, impeller geometry, diameter, tip speed, and shear rate) as well as microscale mixing parameters (those related to power per unit volume, velocity fluctuations, etc.).

C. IMPELLER TYPES AND GEOMETRIES

Having discussed the basic concepts of fluid mechanics applicable to all impellers, we will present various practical impeller types used in industry today. The two basic types of impellers are radial and axial. Radial devices discharge fluid horizontally away from the impeller blades toward the tank wall. Axial flow impellers create flow vertically up or down away from the blades, parallel to the shaft. They produce more flow per unit power than do radial impellers and are more cost effective in flow controlled operations (i.e., solids suspension, blending, heat transfer, some mass transfer applications, etc).

Keeping in mind that a mixer is essentially a pump (albeit not a very efficient one), impeller power consumption creates flow and head. All the power that a mixer supplies to a fluid produces flow, head, or shear. At contant power, the relative amount of flow or shear applied to a fluid can be altered by changing the design of the impeller. Every impeller design creates some balance of flow and shear. There are no pure axial or radial flow impellers, although certain designs are predominantly axial or radial flow. A wide range of flow/shear ratios can be achieved by using different impellers. Fig. 1-3 shows the flow/head (shear) balance for common impeller types at constant power.

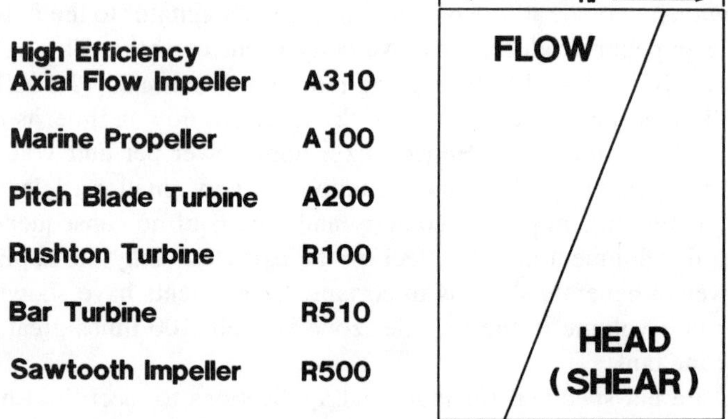

FIGURE 1-3. Impeller spectrum relating flow/head (shear) split at constant power for various types of impellers. (*Mixing Equipment, Inc.*)

Once the specific process requirements are known (i.e., whether the mixing application is flow or head (shear) controlling), the correct impeller type can be selected as the starting point in the design of the mixer.

At the top of the impeller spectrum are impellers with high flow and low shear rates. These are paddle types, large diameter gates, rakes, spirals, and anchors running close clearance to tank walls for high viscosity blending and non-Newtonian slurry suspension applications.

Next are axial flow impellers, including marine propellers, hydrofoils (variable pitch curved blade turbines), and fixed pitch flat blade axial flow turbines. These impellers are generally used for low to moderate viscosity blending, solid suspension, and heat transfer—processes requiring high pumping efficiencies.

For process results requiring progressively higher power per unit volume and/or more shear (head), the family of radial flow impellers becomes more practical. Flat bladed radial impellers, with blades at 90° to the impeller centerline, are often used for gas dispersion and mass transfer applications because of their well balanced flow/shear development.

Extremely high shear rates may be obtained (with relatively low flow production) by using narrow bladed bar turbines, sawtooth impellers, and homogenizers/colloid mills. These high shear impellers are used effectively for high shear dispersions of solids in liquids or liquids in liquids, as well as applications requiring solid size reduction in addition to dispersion (i.e., solids attritioning and dispersion).

Let us examine in closer detail the characteristics of this impeller spectrum. Fig. 1-4 shows the physical appearance and typical dimensions of those basic impellers.

The impeller in Fig. 1-4(a) has a disk and flat blades at 90° to the disk, and is known as the flat bladed radial flow or Rushton turbine. It is used for applications requiring balanced flow/head development with fairly high shear and turbulence formation. This turbine is particularly well suited for gas-liquid contacting because of its flat circular disk. When gas is introduced through a sparger below the impeller, the disk prevents the gas from bypassing the blades or flooding the turbine, instead forcing it on a path to the high shear zone along the lower blade edge.

Power response can be changed by varying the number of bolted blades on the predrilled disk. This measure provides flexibility in performance for any process changes.

The impeller shown in Fig. 1-4(b) is known as a bar turbine. Blades are normally welded in an alternating manner to the top and bottom of the disk as shown. Blade width and height are normally $^1/_{20}$ of impeller diameter. Of all the turbines shown in Fig. 1-4, this one produces the highest shear rate. It runs at relatively high speeds, requiring lower torque, and as such, a smaller speed reduction drive train is required than other radial flow turbines shown.

Radial

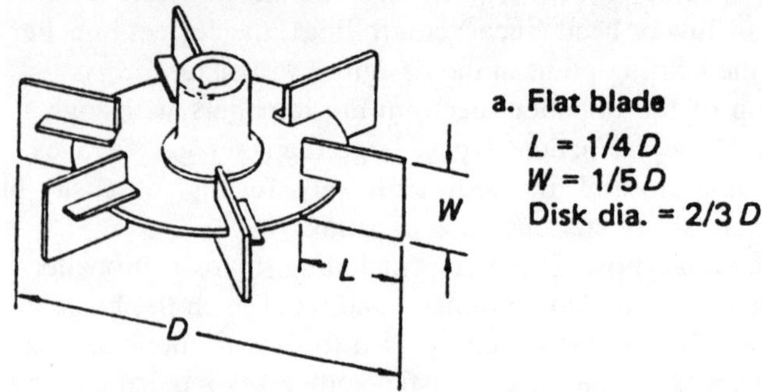

a. Flat blade
$L = 1/4\,D$
$W = 1/5\,D$
Disk dia. = $2/3\,D$

Vertical blades bolted to support disk

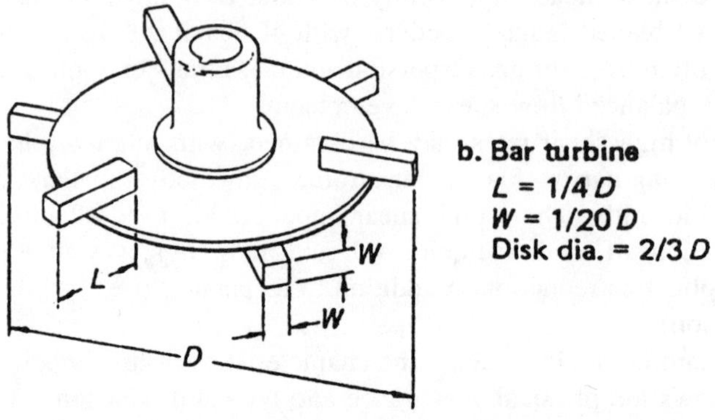

b. Bar turbine
$L = 1/4\,D$
$W = 1/20\,D$
Disk dia. = $2/3\,D$

Six blades—bolted/welded to top and bottom of support disk

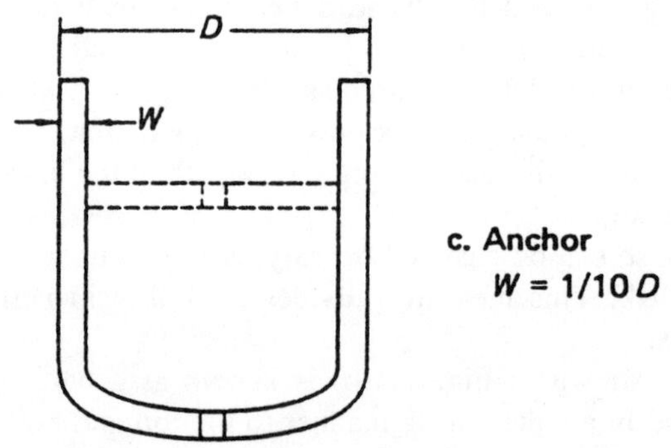

c. Anchor
$W = 1/10\,D$

Two blades with or without cross-arm

FIGURE 1-4(a–g). Impellers commonly found in process industries applications. (*Mixing Equipment, Inc.*)

Axial

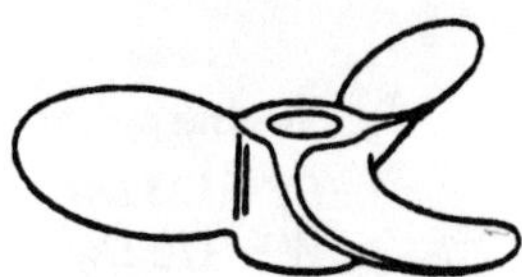

d. Marine propeller, 3 blades
1.5 pitch ratio

Constant pitch, skewed-back blades

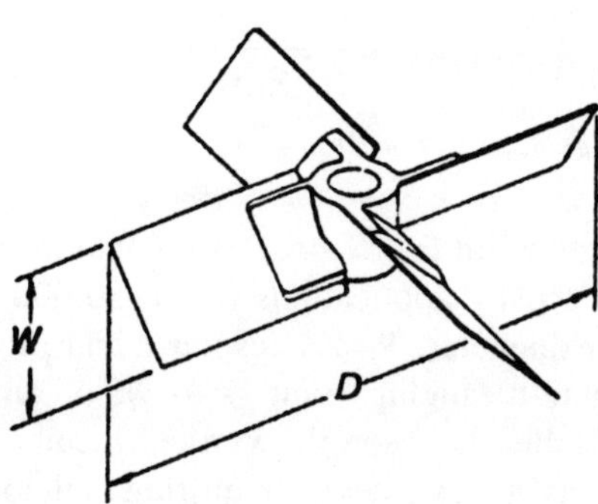

e. Axial flow, 4 blades
$W = 1/5\,D$
45 deg.

Constant angle at 45 deg.

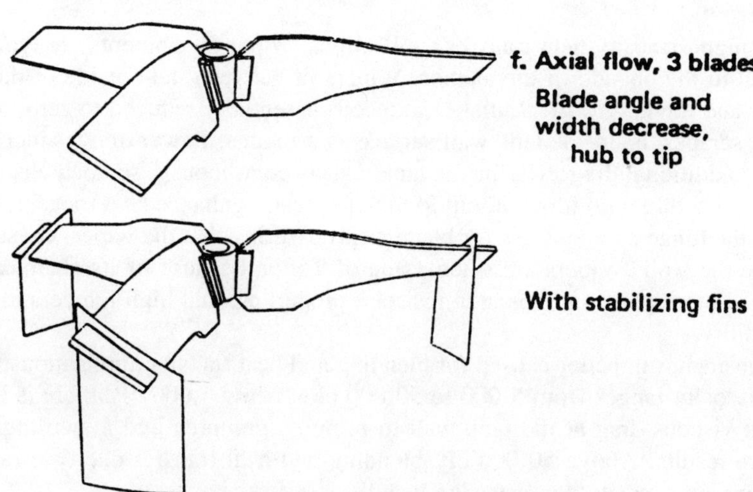

f. Axial flow, 3 blades
Blade angle and width decrease, hub to tip

With stabilizing fins

Variable blade-angle, near-constant pitch

FIGURE 1-4(a–g). (*Continued*)

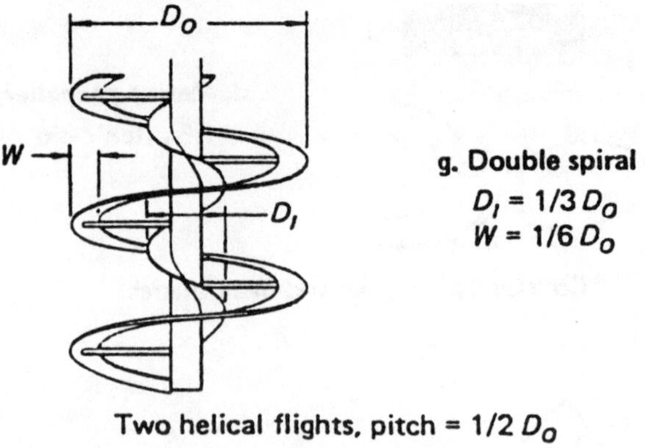

FIGURE 1-4(a–g). (*Continued*)

The anchor or horseshoe impeller, shown in Fig. 1-4(c), is a contoured two bladed radial flow turbine for higher viscosity applications (up to 50,000 cP). Blade width is typically 1/10 of impeller diameter. Power response and process results change very little with blade width ranging from 1/8 to 1/12 of turbine diameter. Very often a crossarm is installed between the vertical members to strengthen the assembly. This turbine runs at slow speeds, requiring high torque and large speed reduction drive in comparison with other radial turbines. The addition of a crossarm has very little effect on power response and mixing result. However, power consumption varies directly with impeller height to diameter ratio.

In high viscosity heat transfer applications, wiper attachments are typically hinged to the outside of the anchor. Wipers fit between the outside of the impeller and the tank wall. Radial clearance is essentially reduced to zero, as the wiper scrapes the inside tank wall surface as it rotates. Power draw is increased by the additional drag. The inside tank wall is continuously scraped clean, resulting in a thin fluid film without fouling to greatly enhance heat transfer. Over time, the hinge compensates for blade wear, assuring that the wipers constantly scrape the wall. Wipers are often made of Teflon because of its chemical inertness, lubricating and abrasion resistant properties, and high temperature applicability.

The anchor impeller is used for blending and heat transfer applications when the viscosity ranges from 5,000 to 50,000 cP. Below 5,000 cP, there is insufficient viscous drag at the tank wall to promote pumping and a swirling flow pattern results. Above 50,000 cP, blending and heat transfer decrease rapidly as pumping capacity declines significantly.

The remaining turbine types (Fig. 1-4(d)-(k)) are axial flow turbines.

The marine propeller, shown in Fig. 1-4(d), is used mainly with portable and fixed mounted mixers ranging from 1/10 to 5 hp. Because the propeller is often cast during manufacture, it is limited in size, typically ranging from 2″ to 36″ diameter. Pitch ratios are 1.0–1.5. This means that, theoretically, if there were no slippage at 100% efficiency, a cylinder of flow would be generated under the propeller, with cylinder height equal to pitch ratio times propeller diameter for each revolution.

The marine propeller produces a downward axial flow pattern. It has a constant blade pitch, created by the twisting skewness of the blades. It is designed to run so that its trailing edge is the one with the smaller radius of curvature. If the marine propeller is run in the reverse direction, pumping efficiency is reduced by 15%.

Mixers mounted on the side of a tank (side-entering agitators) are often installed with marine propellers. They are mounted with the impeller shaft at an angle (typically 10°) to the tank centerline to promote top to bottom turnover of tank contents without swirling. Side entry mixers can be mounted on manway covers on the side of a tank. As such, the marine propellers must be small enough in diameter to pass through the manway flange opening.

Fig. 1-4(e) shows a four bladed pitched blade (45°) axial flow turbine. Blade widths typically range from 1/6 to 1/5 turbine diameter. The number of blades can range from 2 to 6 and blade angle can vary from 30 to 60° to the horizontal. Power response and pumping vary with number of blades, blade angle, and width. The optimum configuration (for maximum pumping) of a four-bladed turbine is 45° angle and 1/5 width to diameter ratio.

Pitched blade turbines are found in larger tanks where high power requirements are needed for flow controlled mixing applications (i.e., low to moderate viscosity blending, solid suspension, and/or heat transfer). Impellers can range from 12 inches to 16 feet diameter with up to 500 hp mixer drives. Marine propellers would be too heavy to be practical in larger tanks. Because of its constant angle, flat plate construction, the pitched blade turbine is relatively inexpensive to fabricate. It may be made as an all-welded construction with blades welded directly onto a removable hub or directly to the shaft. This is by far the least expensive mode of fabrication. An alternative construction mode is to bolt the blades to the hub. Bolted blade construction is preferred to easier installation, and in cases where the tank has a closed top and/or small openings to insert the turbine assembly it is the only practical choice.

If coatings are applied to the turbine (i.e., rubber, urethane, polymers, etc.) for abrasion and/or corrosion resistance, it is recommended that all-welded construction be followed. Coating of blade hardware (nuts, bolts, lockwire) is not recommended, unless the hardware is countersunk below the surface, to ensure good mechanical bond of coating to surface. This recommendation applies to every turbine type, if the impeller is to be coated.

The turbine shown in Fig. 1-4(f) provides the greatest flow and lowest shear of any impeller, radial, or axial flow. The constant pitch, variable blade angle, axial flow turbine has its steepest angle and widest blade close to the hub. Progressing from hub to tip, the angle becomes smaller and blade width narrower. This unique combination of blade twist and variable angle and width provides for a constant pitch ratio across the blade. Thus uniform velocity profile is produced across its entire discharge area, resulting in maximum flow and minimum shear rate.

The constant pitch axial flow turbine can have three to six blades, depending on specific mixing requirements. Low viscosity (less than 2,000 cP) flow controlled applications are best suited for three-bladed turbines. Somewhat higher viscosity applications (2,000–5,000 cP) and gas-liquid applications requiring dispersion of gas and pumping of fluid are best handled by four to six blades with blade widths greatly increased. This blade arrangement compensates for the increased resistance to flow produced by higher viscosity and/or gas introduction.

The constant pitch turbine is typically offered in 20–120 inch diameter range with agitator drives up to 500 hp. Large flow-directing stabilizer fins as shown in Fig. 1-4(f) improve pumping capacity in higher viscosity regimes and in situations where there is low liquid coverage above the impeller (less than ½ turbine diameter of coverage). In addition, the stabilizer fins prevent excessive shaft whipping and deflection in a low liquid coverage scenario.

This turbine performs like a marine propeller, but overcomes that impeller's disadvantages:

1. It weighs less. It is not cast constructed like the marine propeller. Thus the shaft design is not hindered by excessive impeller weight.
2. It is less expensive.
3. It is easily split into three or more sections, bolted together at the hub, for use in tanks with small openings.
4. It may be made of composite materials for corrosive and/or erosive environments, with further reduction in weight and cost.

An improved version of Fig. 1-4(f) is the hydrofoil or fluidfoil impeller. It has all the features of the constant pitch axial flow turbine, except that the blades are curved in addition to twisted, and taper more in width from hub to tip. The hydrofoil provides the highest flow and lowest shear rate for a given power level of any available commercial impeller. Blade profiles are patterned after an ideal fluidfoil section. Blade tips known as proplets effectively eliminate any tendency for the flow to recirculate around the blade tips. Proplets improve flow efficiency by 10% with a corresponding reduction in velocity head by the same amount.

Figure 1-4(g) shows a double spiral impeller for axial flow in high viscosity applications. This impeller has an inner flight or helix pumping downward, while the outer flight pumps upward. The diameter of the inner flight is one-third the impeller diameter. The outer ribbon width is one-sixth the diameter. Impeller height is equal to diameter. Typical sizes range from 20 to 120 inches diameter, at 1–250 Hp, and operates at speeds of 5.5–45 rpm.

The hydrofoil turbine shown in Fig. 1-4(h) (less proplets) was designed using laser analytical techniques to provide maximum flow or pumping capacity per unit of expended power, while at the same time providing uniform velocity profile (constant shear rate) across the full diameter. Because of its unique design, similar flow can be developed for 60% of the horsepower of a fixed pitch axial flow turbine (Fig. 1-4(e)).

Reynolds number is the dimensionless variable defined as the ratio of inertial to viscous forces in the flow regime created by the impeller.

$$N_{Re} = \frac{DV\rho}{\mu} \tag{15}$$

where: N_{Re} = Reynolds number
D = Turbine diameter
V = Velocity developed by impeller

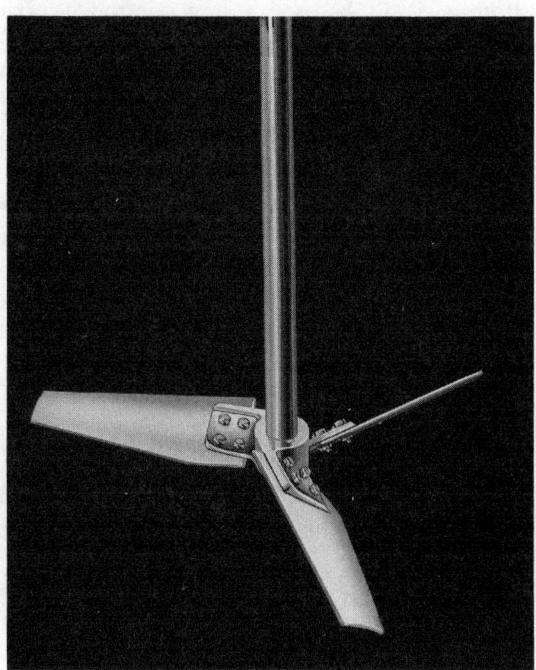

FIGURE 1-4(h). A-310 Fluidfoil impeller, narrow bladed hydrofoil. (*Mixing Equipment, Inc.*)

ρ = Fluid Density
μ = Fluid Viscosity

but since

$$V = ND \tag{16}$$

where N = Turbine Speed, the Reynolds number can be alternately defined as:

$$N_{Re} = \frac{D^2 N \rho}{\mu} \tag{17}$$

The narrow bladed hydrofoil loses flow efficiency in higher head applications. At a Reynolds number of roughly 200, the discharge flow pattern becomes more radial. The low solidity ratio of the narrow bladed hydrofoil reduces its effectiveness measurably. Solidity ratio is defined as the ratio of blade area to the area of a circle with diameter equal to turbine diameter. In fact, a conventional fixed pitch axial flow turbine is more effective at low Reynolds number for overall circulation and for use in higher viscosity flow controlled applications.

As a rule, hydrofoils should be used in flow controlled mixing applications at Reynolds numbers above 1000. At Reynolds numbers below 1000, fixed pitch axial flow turbines are more efficient.

At Reynolds numbers in the range of 1,000–10,000, non-Newtonian fluids require a higher solidity ratio hydrofoil for more effective mixing results. Non-Newtonian fluids exhibit variable viscosity as a function of shear rate, and their behavior may exhibit a time dependancy of viscosity. Newtonian fluids, water for example, have constant viscosity at all shear rates, and do not exhibit time dependency of viscosity. Newtonian fluids are best mixed with narrow bladed hydrofoils, unless the Reynolds number is less than approximately 1000.

For gas dispersion applications, low solidity hydrofoils and axial flow turbines tend to flood (i.e., bypass gas through the blades of the turbine rather than disperse the gas) at relatively low gas rates when compared to the more traditional flat bladed radial flow turbines equipped with a disk (Fig. 1-4(a)). This has led to the development of the wide bladed, high solidity hydrofoil shown in Fig. 1-4(i). Solidity ratios approaching 90% have been accomplished with this design. In the area around the hub there is some overlap between blades.

The difference between fixed pitch impellers and hydrofoils (narrow and wide bladed versions) is explained through the use of velocity vector diagrams. The fixed pitch turbine (Fig. 1-4(j)) has constant blade angle (pitch) across the entire blade length with increased angle of attack, α. Angle of attack is that angle formed by the resultant axial and radial velocity vectors with the impeller pitch.

The hydrofoil (Fig. 1-4(k)) provides a constant angle of attack across its

FIGURE 1-4(i). Typical impeller for gas–liquid dispersion, wide bladed hydrofoil of high solidity (solidity ratio 0.85). (*Mixing Equipment, Inc.*)

blade length. The pitch is steepest at the hub and due to blade twist and curvature becomes shallower toward the tip. The hydrofoil is a variable pitch, constant angle of attack impeller. Fluid lift, synomymous with pumping efficiency, is constant across the blade of a hydrofoil. Lift is variable along the blade of any fixed pitch turbine. The hydrofoil not only develops uniform lift, but maximizes it as well if the angle of attack is designed specifically for the application at hand. This approach gives greater flow and better mixing.

The streamline contour of the hydrofoil blades prolong their life expectancy in abrasive and erosive mixing environments. Compared to fixed pitch turbines, hydrofoils are not machine-shop reproducible. Special fixtures, jigs, and presses are required to produce the proper blade curvature and twist. This equipment lies solely with the original equipment manufacturers. Because of their sophisticated design and manufacturing requirements, hydrofoils are more expensive than conventional fixed pitch turbines. High solidity hydrofoils may weigh more than their fixed pitch counterparts, requiring larger shaft diameters. However, these negatives are usually more than offset by the advantages of the hydrofoil extended blade life in severe duty services, improved process results, and significant power savings (up to 40–50%).

Hydrofoils are not all the same in design and performance. A number of

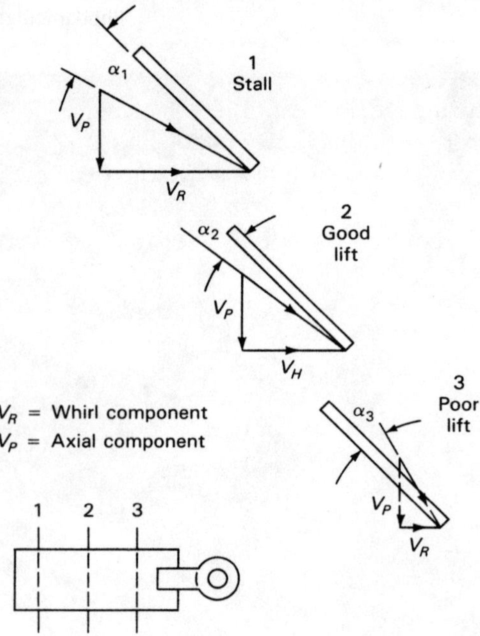

FIGURE 1-4(j). Fixed pitch impeller, variable angle of attack. (*Prochem Mixing Equipment, Inc.*)

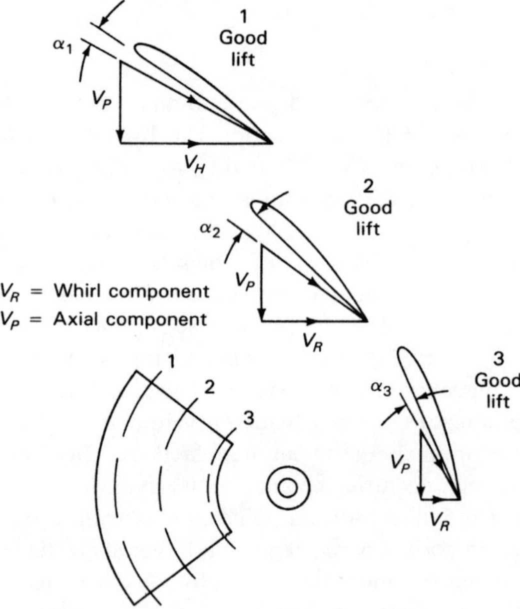

FIGURE 1-4(k). Variable pitch hydrofoil, constant angle of attack. (*Prochem Mixing Equipment, Inc.*)

variables must be taken into account in coming up with the proper design to meet the specific mixing requirement. These variables include diameter, number of blades, camber, pitch, blade area or solidity ratio, hub size, and speed. Process conditions can change such that a hydrofoil designed to be most efficient in a low viscosity fluid (i.e., a narrow bladed trifoil) can stall, or stop pumping, in higher head applications such as gas dispersions, increased viscosity, or other process changes and upsets. As head increases, the angle of attack increases due to reduced axial component of velocity, V_p, in Fig. 1-4(j) and (k). Compensation for this increased head created by process changes and/or upsets is best addressed by a hydrofoil design of more blades (more than three blades), larger hubs, greater blade solidity (area), and proper blade pitch. Hydrofoils of this design are more forgiving to process changes and/or upsets.

Hydrofoils must be prudently designed not only for normal operating conditions, but also for conditions anticipated during process upsets and/or changes. It is impractical to replace hydrofoil impellers every time operating conditions change. A more practical approach is to design and install a hydrofoil that is somewhat inefficient, but still functional and productive, under upset conditions, while of optimum design and efficiency under normal operating conditions. So-called standard, off-the-shelf, hydrofoil designs are optimum for a limited application and operation spectrum. Very often the design must be one of an application/operation specific approach to optimize mixing results.

Less efficient axial and radial flow turbines may be retrofitted with more efficient hydrofoils to achieve improved process results and/or reduce agitation power requirements. Retrofitting with hydrofoils should be pursued with the suppliers of these impellers, since a thorough process and mechanical analysis must be conducted. A check must be made, for example, of the drive train and shaft design to ensure that increased thrust and forces with the hydrofoil will not jeopardize the mechanical integrity of the mixer. Retrofitting will be discussed in more detail in Chapter 6.

D. IDEAL TANK SHAPES AND FLOW PATTERNS

The fluid force regime in an agitated vessel arises from the rotation of the impeller, is impressed upon the fluid in the tank and is modified and ultimately changed by the static vessel wall. The influence of the tank on the fluid flow regime, and therefore the mixing, is discussed here.

Vertical cylindrical tanks with height equal to diameter are ideal geometric proportions for top-entry agitators. Short, squat tanks with height less than diameter very often require multiple top-entry machines to assure mixing in all zones of the tank. In fact, tanks with height less than diameter are often better served by side-entry mixer(s) to achieve desired mixing results for less installed cost.

Tall, thin tanks with height greater than diameter require a single top-entry mixer with multiple turbines. More than one impeller is required to assure mixing in all vertical zones. Each impeller on the shaft acts like a pump in series with the others. Flow is additive while overall head development remains nearly constant. The adjacent turbines above and below each impeller act as resistances to flow development by that impeller. As a rule, multiple turbines of equal diameter add 70–80% of full theoretical flow, with the percentage decreasing as the number of turbines increases.

Compared with side-entry devices at a given power rating, top-entry mixers operate at lower speeds with larger impellers. Pumping rates are higher, resulting in shorter blend times. Speeds range from 30 to 100 rpm, compared with 280 to 420 rpm for side-entry units.

Since top-entry mixers produce the highest flow at constant power, they are used more often than side-entry ones. However, each type of machine has its own advantages and limitations in given tank geometries as previously discussed. Top-entry agitators run at higher torque than side-entry units, resulting in a more expensive mixer at a given power level. However, offsetting the higher capital cost of the top-entry machine is its lower operating cost (power) to achieve the same mixing result at constant flow. As a general rule of thumb, a top-entry agitator cost twice as much as a side-entry unit, but requires half the power to accomplish the same flow in similar tank geometries.

Multiple agitator mountings (regardless of whether they are in top- or side-entry orientations) are often required on large diameter tanks, with height much less than tank diameter, i.e.,

$$H/D \leq .5$$

where: H = tank height
D = tank diameter.

As with multiple turbines on a common mixer shaft, adjacent mixers act as resistances to flow development by each agitator. Multiple agitators of common design add 70–80% of full theoretical flow, with the percentage decreasing as the number of multiple mixers increases.

Axial- and radial-flow impellers are used for top-entry mixers. For axials, fluid is circulated through a single loop in the high turbulence zone of the impeller by a flow parallel to the shaft axis. This type of action provides repetitive circulation through the zone, while allowing blending to occur in a single flow pattern. Much like a pump, an agitator turbine in rotation has a suction and discharge zone with pressure gradients across the impeller creating the driving force for pumping.

Radial flow impellers, on the other hand, have two suction zones and one discharge zone, producing a dual or split flow pattern. One flow path is created above and a duplicate flow pattern below the turbine centerline. Since radial flow impellers produce less pumping than axial flow turbines at similar power levels, blend times are longer with the former type.

Significantly greater power is required for radial flow turbines compared with marine propellers to achieve similar flow rates and equal blending results. The magnitude of the difference depends upon efficiency differences. Regardless of impeller type, tank baffles are required to establish proper flow pattern with top-entry machine(s) of vertical shaft alignment (i.e., no angled entry of mixer shaft into the tank).

Top-entry mixers can be mounted in a number of different ways, including on center, off center, and angular. The wide variety of mounting fixtures, shafts, and impellers allows for operation above and below the first natural frequency of the mixer shaft—the speed at which harmonic vibrations occur. Often, portable mixers are supplied with C-clamps for moutning to the top rim of an open top tank. Portables are often mounted with ball and socket joints for angular entry, thus avoiding the need for baffles in the mixing tank.

Side-entry units are best suited for blending homogeneous fluids or slurries with low-settling-velocity solids.

The most common applications are for oil and gasoline blending and storage tanks, paper stock blend tanks, and flue gas desulfurization units. A disadvantage of side-entry mixers is the need for a stuffing box or mechanical seal to prevent in-tank fluids from leaking through to the outside along the mixer shaft. The seals will leak unless properly maintained and operated. Side-entry agitators of up to 50 hp can be supported directly from the mounting nozzle and tank wall. For larger units, up to several hundred horsepower, independent support bases are used. Very often side-entry machines are mounted on a manway cover large enough to allow passage of the propeller/impeller still on the shaft.

Establishing the correct flow pattern with side-entry agitators is essential. The optimum configuration is shown in Fig. 1-5. A side view appears Fig. 1-6. Note that side-entry agitators do not require tank baffles to achieve proper flow patterns.

The configuration shown in Fig. 1-7 must be used too. This ensures that the impeller is mounted in the correct direction for its rotation. If the proper flow pattern is not followed, tangential swirl will result, and top-to-bottom turnover will be poor. Mixing will be nonuniform and blend time greatly increased.

Top-entry mixers mounted on the tank centerline require baffles to assure good mixing and proper flow pattern throughout the entire vessel. Baffles convert tangential fluid motion induced by the impeller rotation to vertical circulation. Fig. 1-8 shows this top-to-bottom flow pattern. Excessive baffling reduces mass flow and localizes the mixing by putting excessive drag on the rotational motion of the fluid. Power draw is increased without benefit of improved mixing. This can result in poor performance.

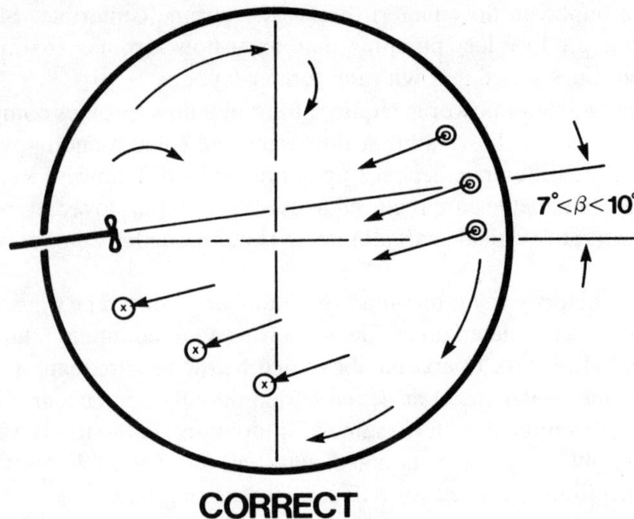

FIGURE 1-5. For side entry mixers, proper impeller placement will ensure good mixing. (*Mixing Equipment, Inc.*)

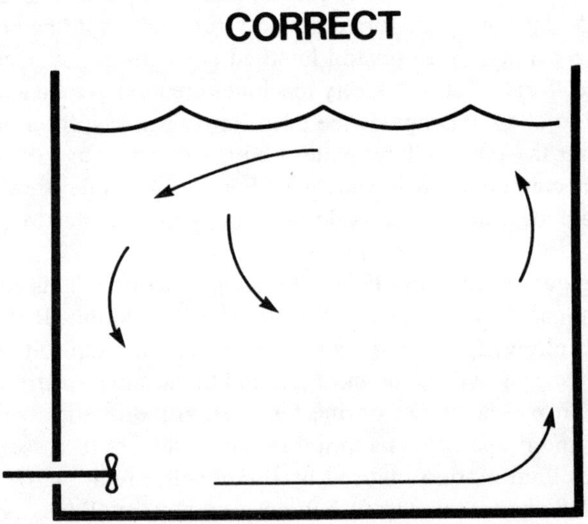

FIGURE 1-6. The correct configuration in Fig. 1-5 gives this flow pattern., viewed from the side. (*Mixing Equipment, Inc.*)

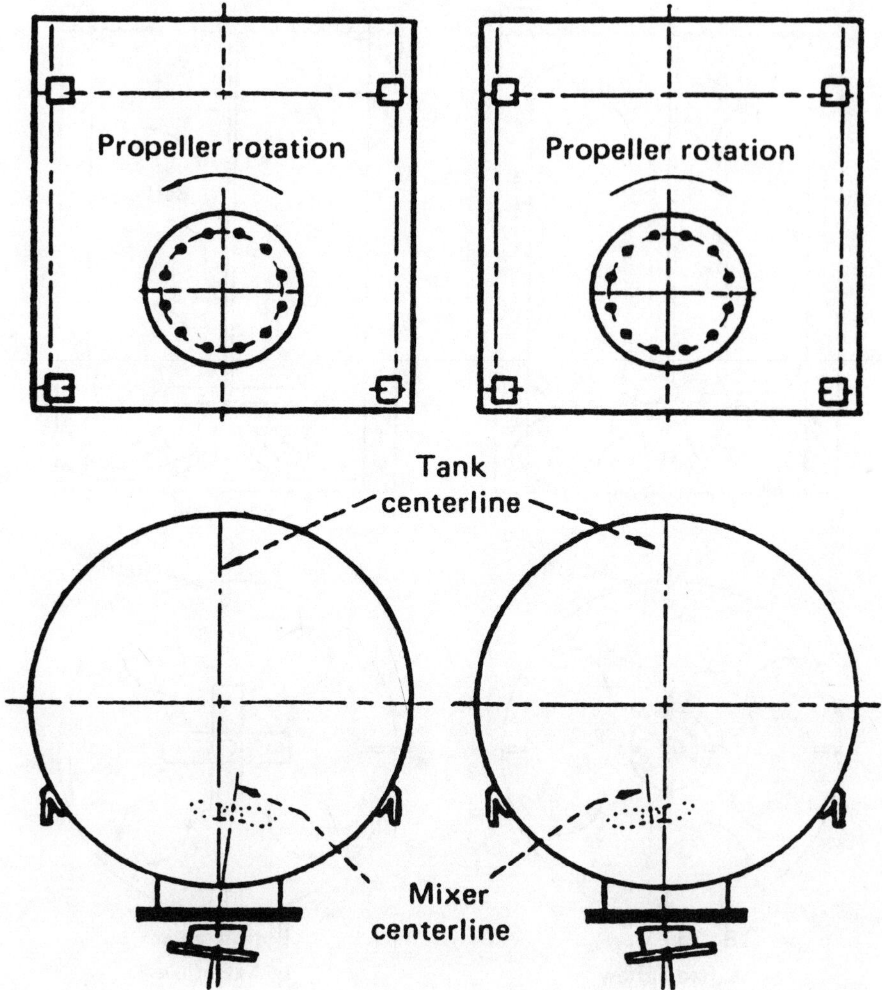

FIGURE 1-7. Side entry mixers installed at proper angle and position for propeller rotation. (*Mixing Equipment, Inc.*)

Baffles must be provided for low-viscosity mixing when top-entry mixers are located on the tank centerline. This includes all types of mixers and impellers/propellers.

Baffles assure that the entire batch will pass through that zone of the impeller where there is maximum agitation. Baffles also:

- Promote the flow pattern required for the process

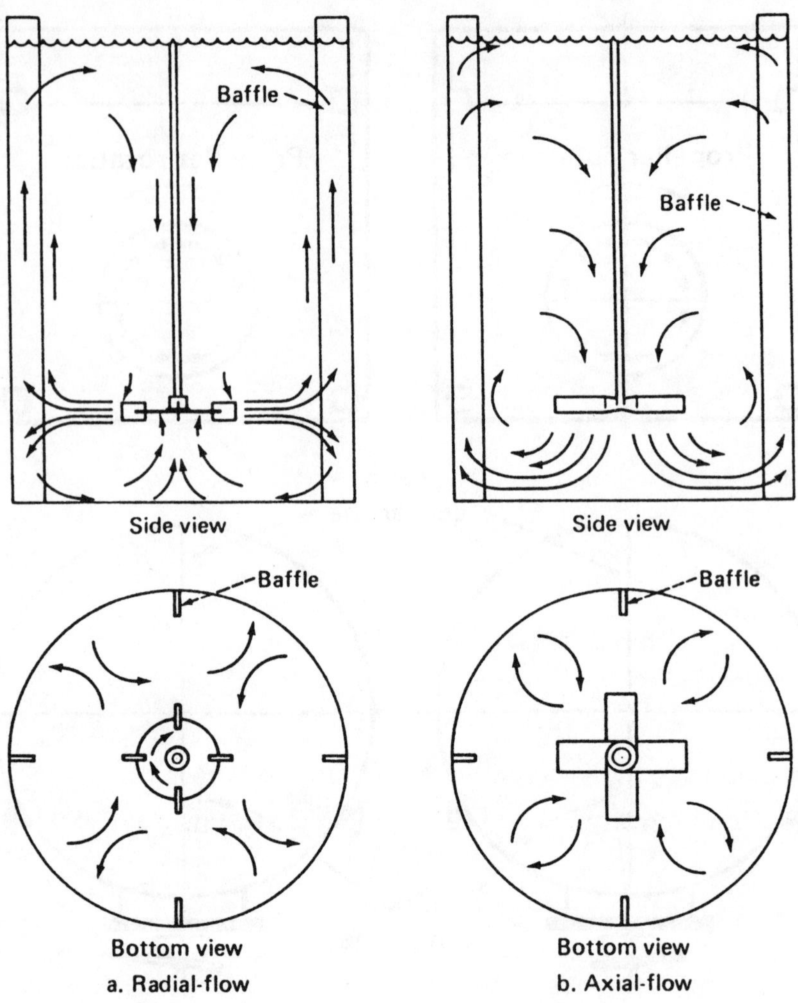

FIGURE 1-8. Flow patterns for radial and axial flow impellers are improved with baffles. (*Mixing Equipment, Inc.*)

- Direct flow from the impeller, producing the required vertical, top-to-bottom currents
- Change the flow from a rotary to a mixing pattern
- Avoid excess swirling, vortexing and air induction
- Greatly improve mixer-loading accuracy. The use of baffles yields more accurate data, resulting in better mixing correlations and predictable process results. Thus mixing loading—specifying any two of the parameters P, N, and D—is improved

- Assure a stable, consistent power draw
- Produce more uniform radial shaft loads, allowing for longer shaft lengths.

Vertical cylindrical tanks should be equipped with four baffles, 1/12 the tank diameter in width, extending vertically along the straight side of the tank and located 90° apart. Wider baffles provide slightly stronger vertical mixing currents, but may act as flow dampeners by reducing mass flow and rotary motion.

Fewer or narrower baffles allow for more rotary motion but also reduce impeller draw. This limits the energy that can be applied to the batch, because of swirling at higher power levels.

When carryover from one batch to the next is undesirable, baffles should be self-cleaning. In waterlike batches, offsetting standard width baffles one-third of the baffle width from the tank wall permits the flow to scour the area behind the baffle. Unless the baffle is offset from the tank wall, the area immediately behind the baffle tends to be a dead spot. For higher viscosity batches, (i.e., $\mu \geq 10{,}000$ centipoise), the offset distance is one-half the baffle width. As a rule, for open top tanks, baffles should be made of the same thickness as the tank wall, but not less than 1/4 inch. Flow induced loads act on baffles requiring proper support structure. These loads are a function of the type of mixing impeller, D, N, hp, etc., and must be calculated by the mixer supplier. Once the supplier provides the loading on the baffles, the user can properly design an adequate baffle support structure.

Baffles can be eliminated for propeller mixers by mounting the units in an angular off-center position as shown in Fig. 1-9. Off-center top-entry mixers

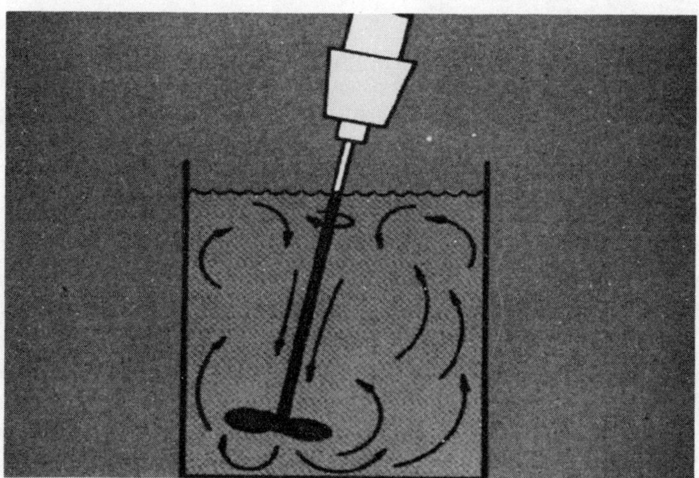

FIGURE 1-9. Angular off-center positioning eliminates the need for baffles in the mixing tank. (*Mixing Equipment, Inc.*)

achieve good top-to-bottom turnover and eliminate swirling. This is particularly effective for non-Newtonian fluids, such as paper stock, starch slurries, etc.

There is a limit to how much power can be applied to a top-entry mixer before vortexing begins, so such configurations are unsuitable for high-power applications. Power levels exceeding 10–15 Hp per 1,000 gallons of tank contents will produce vortexing, even in a baffled tank, in low viscosity fluids. Also, as the shaft rotates, an unbalanced unidirectional fluid force develops on the impeller. This causes stress reversal on the shaft, and the shaft diameter must be made larger as a result. As the shaft rotates, it goes through compression and tension on each revolution, subjecting it to fatigue.

Vortexing flow patterns will occur with top-entry mixers, mounted on the tank centerline, without baffles in waterlike viscosities. Vortexing will also occur in baffled tanks with highly powered machines. Some process results are best suited for vortexing flow regimes. These include certain slurry makedown applications with difficult to wet out solids (i.e., clays, talcs, starches, etc.), gas liquid mass transfer applications, where induction of the gas from above the liquid level back into the liquid via vortexing will enhance the mass transfer.

Portable or fixed mount top-entry agitators, with or without stuffing boxes or mechanical seals, still allow for operating above or below the first natural frequency. However, for a given shaft diameter such as setup will not allow for the use of all shaft lengths.

Larger mixers employing radial and axial flow turbines are usually mounted on rigid support bases. These structures can be beams spanning the top of tanks, mounting flanges, or combinations of the two. Axial and radial flow impellers are typically larger in diameter than propellers and hence run at slower speeds in larger tanks. These impellers are more stable mechanically, but do require a larger motor.

A single radial flow turbine will produce two distinct mixing zones in a tank, as shown in Fig. 1-8. Multiple radial flow impellers on a single shaft will produce a greater number of zones or stages in the tank. This type of flow distribution is beneficial to plug flow reactors or to promote holdup of reactants (particularly gas) in a specific zone. Zone agitation leads to nonuniform distribution of reactants and temperature gradients throughout the stirred vessel. In a continuous flow scheme in a baffled tank being mixed with multiple radial flow turbines, where some degree of control from inlet to outlet is required, this configuration can be used along with proper location of the inlet and outlet to provide control of residence time and reduce short circuiting. Axial and radial flow patterns in baffled and unbaffled tanks are shown in Figs. 1-10 and 1-11.

Typical radial flow impellers are shown in Fig. 1-12. The radial turbine is the most common type, while the chevron and high shear turbines are specialized types of radial turbines. The chevron impeller is most often used to provide agitation in a cone bottom tank. The high shear turbine is designed to produce

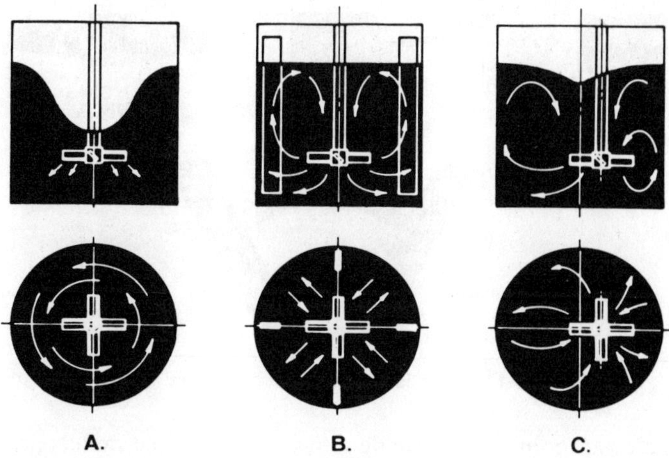

FIGURE 1-10. Axial turbine flow patterns. (*Eastern Mixers, Inc.*)

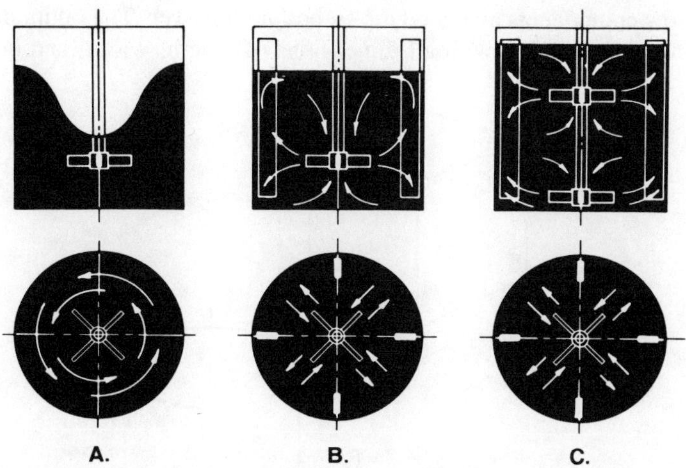

FIGURE 1-11. Radial turbine flow patterns. (*Eastern Mixers, Inc.*)

higher shear than a standard radial turbine and is often used in various dispersion applications.

As previously mentioned, tank shapes and geometry affect the design of the agitator to achieve the desired mixing result. Cylindrical tanks, regardless of whether they are horizontal or vertical in configuration, are of optimum geometry when the diameter and height (or length) are the same. Cost effective mixing occurs in tanks of these geometries. Power requirements and capital cost

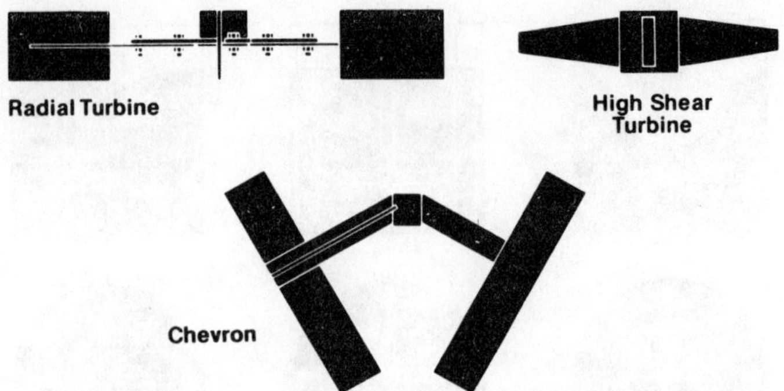

FIGURE 1-12. Radial flow impellers. (*Eastern Mixers, Inc.*)

of the agitator are minimized. Single rather than multiple mixers are required, with only one turbine in lieu of multiple turbines, on the agitator shaft.

Square or rectangular shaped tanks do not require baffles to promote a good mixed flow pattern. In essence, the tank walls act as baffles in providing for vertical flow components and good top-to-bottom turnover. The optimum square tank geometry is a cube, with all dimensions of length, width, and height the

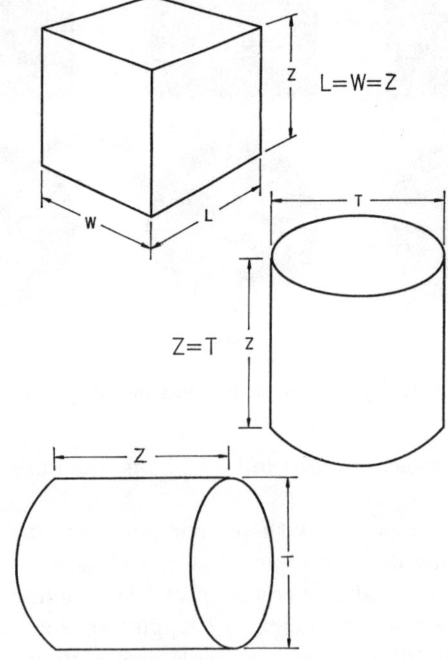

FIGURE 1-13. Cost effective tank geometries for agitator selection. (*Prochem Mixing Equipment, Inc.*)

same. This shape provides for the most cost effective mixer design. Multiple mixers are required in rectangular shaped tanks with tank length greater than width and height. As a general rule of thumb, as tank length approaches some multiple of width, the same multiple of agitators is required to assure good mixing in all zones of the rectangular tank. The corners of rectangular tanks are poorly mixed, and this tank geometry should be avoided particularly for solid suspension and high viscosity application.

For vertical cylindrical tanks with height greater than tank diameter, multiple turbines on the mixer shaft become a requirement to assure proper mixing at all depths of the liquid. A general rule of thumb calls for dual radial flow turbines with a tank height to diameter ratio approaching 1.2–1.5, triple turbines with a ratio approaching 2.0–2.5, etc. High flow efficiency axial flow turbines (i.e., hydrofoils) can be spaced further apart in tall cylindrical tanks. The general rule

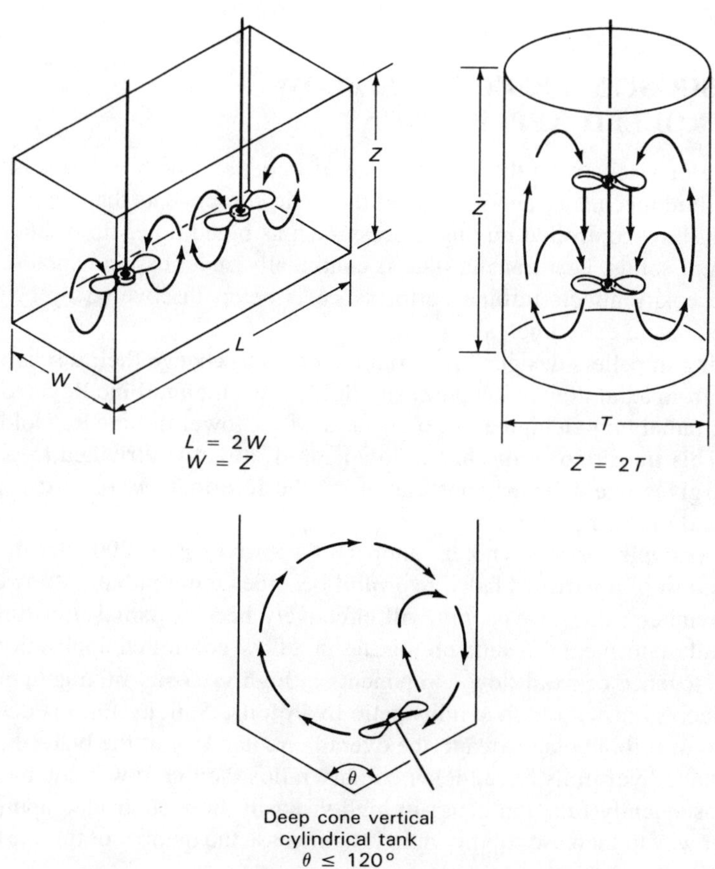

FIGURE 1-14. Nonideal tank shapes with optimum mixer installations.

calls for dual axial flow turbines with tank height to diameter ratios approaching 2.0, triple turbines with height to diameter ratios approaching 3.0, etc.

Ideal tank shapes and geometries are shown in Fig. 1-13. Nonideal tank shapes with multiple turbine requirements and nonconventional mounting arrangements are shown in Fig. 1-14.

Deep cone bottom tank (with included cone angle $\theta \leq 120°$) require bottom-entry mixers, such as that shown in Fig. 1-14, to provide proper top-to-bottom flow patterns. The mixer may be mounted on the side of the cone, or on the very bottom of the cone, rotating in such a manner and with a turbine blade pitch commensurate with pumping up away from the cone bottom. This will ensure good sweeping action of flow across the cone bottom.

It is not recommended to install a top-entry mixer on a deep cone ($\theta \leq 120°$) bottom tank, particularly one mounted directly on center. The flow pattern from such an agitator mounting creates a dead zone of mixing in the very bottom of the cone.

E. VISCOSITY EFFECTS ON FLOW CONTROLLED APPLICATIONS

The role of viscosity in the performance of impellers, both in terms of fundamental fluid mechanics and in some of the basic components that are of concern with any flow controlled mixing process such as blend time, flow pattern, suspension of solids, heat transfer, etc. is continually under experimental scrutiny. How viscosity affects turbine performance is being discovered by better and more extensive experimentation.

Mixing impellers designed for axial flow tend to change their discharge flow pattern from axial flow in low viscosity fluids (i.e., high turbine Reynolds numbers) to radial flow in higher viscosity fluids (i.e., lower turbine Reynolds numbers). This transition of discharge flow pattern, directly attributed to viscosity effects, gives one a better appreciation of the role that viscosity can play in mixer and process performance.

For example, at a Reynolds number of approximately 200, the discharge flow pattern of a narrow bladed hydrofoil becomes more radial. At lower Reynolds numbers, the flow pattern will effectively become radial in nature, with potentially detrimental results on mixing in a flow controlled application.

Maintenance of axial flow components in high viscosity mixing application can be achieved with high-solidity-ratio hydrofoils. Solidity ratio is defined as the ratio of turbine blade area to the overall circular area of the blade diameter. Wide blade hydrofoils have higher solidity ratios then narrow blade hydrofoils and consequently function better in high viscosity flow controlled application. Another way to increase solidity ratio is to increase the number of turbine blades, in addition to making them wider.

In very high viscosity flow controlled mixing applications (i.e. $N_{Re} < 10$), helical or anchor type impellers with close blade clearance to the tank wall will be required.

There is a significant difference in the characteristics of helical or anchor impellers and turbine types. They normally require about 1/5 to 1/10 the horsepower of a turbine impeller for the same overall circulation pattern in high viscosity service. While the axial flow turbine appears efficient, the microscale energy dissipation rate is much higher for turbine impellers than anchors and helical impellers in high viscosity, flow controlled processes.

The anchor and helix impeller are most efficient in high viscosity (Reynolds number less than 10) liquids. Above Reynolds number of 10, these turbines begin to lose efficiency in comparison with axial flow turbines. They produce flow by displacement with very little shear and produce bulk mixing. The anchor and helix impellers are shown in Fig. 1-15.

Classical fluid mechanics analysis is particularly appropriate where high viscosity is present. The fluid flow patterns are around various shapes and objects (i.e., baffles, internal piping, etc.), and the general nature of the fluid boundary layer is somewhat more predictable than when turbulent flow is developed.

As previously stated, narrow blade hydrofoils will have few advantages below a Reynolds number of 200. Here an axial flow turbine of constant blade pitch (i.e., 45° pitched blade axial flow turbine) is generally preferred, down to, say, a Reynolds number of 10. Anchors and helices become the impeller of choice at high viscosities (Reynolds number below 10).

Most high viscosity liquids are pseudoplastic, their viscosities decreasing with increasing shear rate, as shown in Fig. 1-16. Also with these liquids the power relationship changes and becomes

$$\text{HP} \propto N^2 D^3 \mu \tag{18}$$

Therefore, it is very important to have viscosity vs shear rate data for the fluid to be mixed. Various types of viscosimeters may be used to generate viscosity and shear rate data. In one common viscosimeter, the Brookfield, various spindles run at various speeds in the fluid providing viscosity and shear rate data

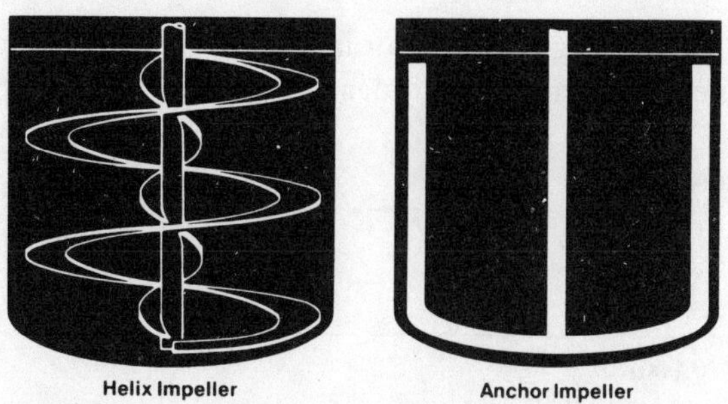

FIGURE 1-15. Specialized impellers. (*Eastern Mixers, Inc.*)

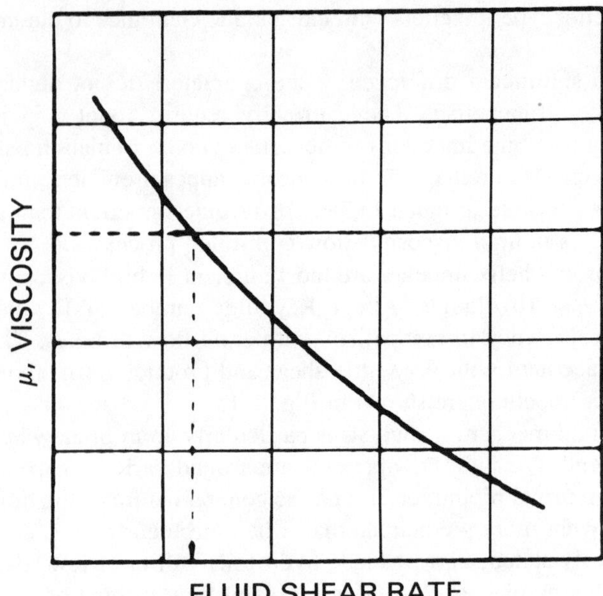

FIGURE 1-16. Typical viscosity versus shear rate plot for pseudoplastic materials. (*Mixing Equipment, Inc.*)

which can be used to properly calculate power draw of the mixer and guarantee proper mixing results.

Viscosity errors have an effect on power drawn by the turbine. In general

$$P \propto \mu^a \tag{19}$$

where: a is listed in Table 1-1,
μ is viscosity,
P is power drawn by the mixing impeller,
N_{Re} is Reynolds number.

In the turbulent flow regime, an error in viscosity value will not greatly affect power draw unless the error is an order of magnitude. However, transitional

TABLE 1-1. Effect of Viscosity on Power.

Flow Regime	Exponent a: $P \propto \mu^a$	
	Top-Entering	Side Entering
Turbulent ($N_{Re} > 10{,}000$)	0	0.16
Transitional ($20 < N_{Re} < 10{,}000$)	0.5	0.8
Laminar ($N_{Re} < 20 - 100$)	1.0	Not recommended

and laminar flow mixing deserve caution and accuracy of measurement of viscosity to estimate the power draw.

Sequential blending and mixing of batches with several components is a likely scenario for viscosity to vary during the batch cycle. Using a single agitator designed for the highest fluid viscosity can result, in such a case, in overdesign or equipment failure. Instead, the use of dual agitators, a two speed motor or variable speed drive train for variable viscosity mixing applications is often the preferred and optimum design.

High viscosity pseudoplastic slurries (non-Newtonian fluids that are "shear rate thinning," i.e., as shear rate increases, viscosity decreases) present a number of problems when one is trying to properly select the agitator and calculate power draw. The change in "apparent viscosity" with shear rate often results in improper power loading and/or agitator design if only a single "viscosity" is considered.

A proven analytical technique for mixer power draw with non-Newtonian fluids and pseudoplastic slurries is as follows: For pseudoplastic slurries, the following relationship applies:

$$\langle \text{Apparent Viscosity} = K(\text{Shear Rate})^{(N-1)} \rangle \tag{20}$$

A log-log plot of apparent viscosity versus shear rate will be a straight line with slope $(N - 1)$.

Impeller viscosity, which is a function of rotationial speed N for a given series of geometrically similar impellers, is the viscosity "seen" by the impeller within its periphery, and at its sheer rate.

Process viscosity, which can be shown to be a function of speed N, impeller diameter to tank diameter ratio D/T, and the slope of the viscosity–shear rate curve $(N - 1)$, is the bulk fluid viscosity in the tank based on average shear rates.

Determination of impeller viscosity will allow one to properly load the mixer powerwise and prevent under- or overloading conditions.

The process viscosity defines the bulk behavior of the system in terms of average sheer rates existing in the geometry of the tank.

Usually, the basic data for determining impeller viscosity and process viscosity are viscosimeter readings. Typical lab data for a slurry of asphalt and grit are as follows. (Brookfield RVF viscosimeter, spindle No.6, Temp. 76°F, specific gravity 1.43):

Spindle, RPM	Reading	Multiplier	Viscosity, cps
2	15.5	5000	77,500
4	17.5	2500	43,750
10	25.5	1000	25,500
20	40.5	500	20,250

The highest and lowest figures of viscosity are plotted in Fig. 1-17, and the line is extended. Impeller viscosity is determined by drawing a vertical line from the full scale impeller type and speed (i.e., axial flow fixed pitch turbine at 68 RPM) and reading the viscosity at the left edge. Its value in the example plot is 9500 cps. This viscosity is used for impeller selection and power draw calculation.

The reference viscosity is the first approximation of the process viscosity. It is read at a spindle speed of 20 RPM. Its value in this plot is 20,250 cps.

As previously discussed, different impellers are used for high viscosity flow controlled mixing applications. The most common types are the anchor and helix (helical or spiral) impellers shown in Fig. 1-15.

Both the anchor and double helix turbines are close clearance impellers, running very close to the tank walls. They have different flow patterns. The anchor does not provide top-to-bottom turnover in viscous mixing requirements. Most of its mixing is done in eddies formed in the wake of the anchor blades. The flow pattern is more radial than axial in nature. The helical impeller, on the other hand, does provide top-to-bottom flow velocities in viscous fluids and provides better mixing results in viscous, flow controlled applications, at similar power draw to the anchor. Optimum top-to-bottom turnovers may be achieved

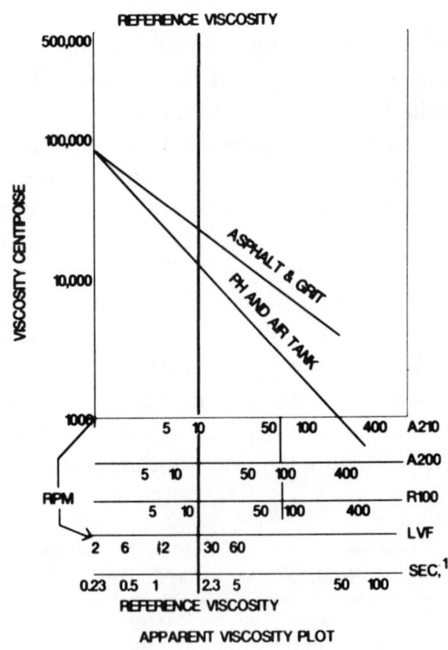

FIGURE 1-17. Apparent Viscosity plot. (*Mixing Equipment, Inc.*)

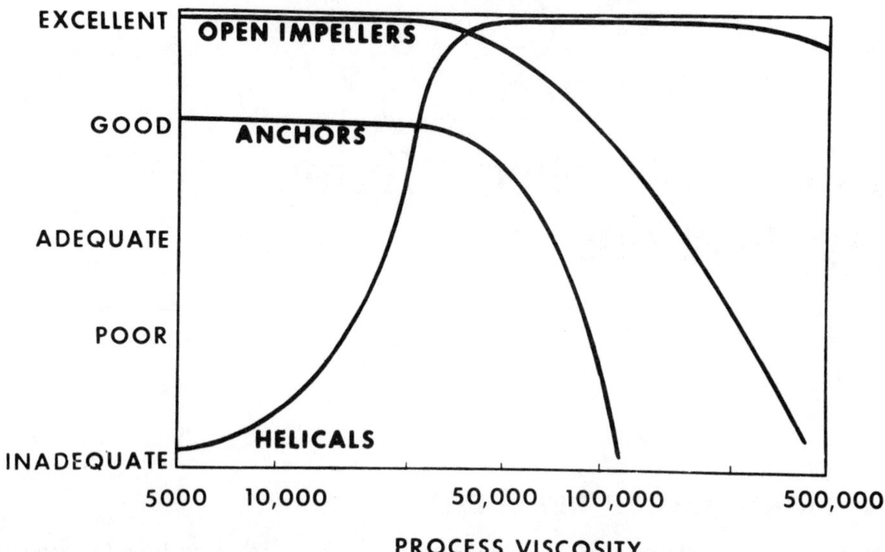

FIGURE 1-18. Relative pumping capabilities of different impellers. (*Mixing Equipment, Inc.*)

with double helical impellers with the flights being the reverse pitch to each other (i.e., the inner flight pumping downward, while the outer one pumps upward or vice versa).

When to use which type of impeller is again a process consideration as well as one of economics. The anchor is less expensive, but does not provide the same degree of mixing as the helical impeller. Therefore greater operating cost (horsepower) with the anchor will be required to achieve similar mixing results, more than compensating for its lower capital cost.

Since both impellers operate at close clearance to the tank wall, their heat transfer capabilities are similar in medium viscosity fluids (20,000—60,000 centipoise). Each process must be evaluated individually to determine which impeller is optimum for the mixing application at hand.

Figure 1-18 shows the comparison of relative pumping capacities of open axial and radial flow turbines to anchors and helical turbines at various viscosities. Helical turbines operate best above 20,000 centipoise, whereas anchors provide good pumping characteristics up to 60,000 centipoise. Open impellers, on the other hand, maintain good pumping capacity up to 100,000 centipoise. As viscosity increases, open impellers must get larger in diameter relative to the tank diameter and/or require multiple impeller configurations to maintain good flow characteristics.

2

Flow Controlled Mixing Applications

A. LIQUID–LIQUID BLENDING

Blending of liquids is described as the operation of combining liquid components of different viscosities and/or densities to produce an end product with uniform properties throughout, such that the end product will not separate into component phases with time.

Blending of miscible liquids is a simple physical mixing consisting of combining two or more materials until the particles, portions or drops of each of the components are disseminated within each other satisfactorily. The degree of mixing or intimacy of the particles is a matter of subjective judgement as to what is acceptable and necessary.

Liquid blending can be classified into two broad classifications:

Low Viscosity—$<$50,000 centipoise
High Viscosity—$>$50,000 centipoise

The delineation is at 50,000 centipoise because different impellers are used in high viscosities versus low viscosities, as we saw in Chapter 1. This is by no means a hard, fast dividing line, but is to be used only as a guideline in determining what types of impellers should be considered.

Regardless of what type of blending application is at hand, blending controls the same variables in any process:

1. Temperature uniformity
2. Reactant–catalyst uniformity
3. Product uniformity
4. Fluctuations

Each particular process will govern how much variation in any one of these variables is acceptable. The degree of blend uniformity needed is most certainly a process consideration.

B. LOW VISCOSITY LIQUID-LIQUID BLENDING

With an understanding of what blending can do processwise, the following factors and their effects on agitator design must be considered.

1. *Characteristics of Material Being Blended.* This includes the viscosities and densities of each component, the percent of each relative to the final product, and the final product viscosity and density. The difference in viscosities and/or densities of the components and final product provide the "driving force" for blending to take place. The greater the differences in viscosities and/or densities of the components, the more horsepower is required. Also, the higher the final product viscosity, the more power is required.

2. *Starting Conditions.* This refers to whether the agitator will be running during addition of the second component, and third component, and so forth. It is always recommended that the agitator operate during addition of components to minimize blend time to achieve uniformity. If the components are allowed to stratify during addition without the mixer running, it may take 5 to 10 times longer to bring the contents to blend uniformity by running the agitator after rather than during component addition.

3. *Blend Time.* Blend time is defined as the time it takes, from when all the components have been added to the tank with the mixer running, until the tank contents are brought to a defined blend uniformity or final product consistency. The specific process requirements will dictate what blend time is needed and to what degree of uniformity the components must be blended.

4. *Tank Size and Geometry.* The larger the tank, the greater the power required to obtain the same blend time. As the tank height (and therefore batch depth) to tank diameter ratio, Z/T, increases, multiple turbines become necessary.

In summary, the following are important considerations influencing mixer selection for liquid-liquid blending applications. They must be answered and quantified to assure proper mixing results and optimum agitator design:

1. What blend time does the process require?
2. What degree of uniformity does the process require?

As previously discussed, agitators for liquid-liquid blending can be used in one of two ways:

On line. Mixers can be used for continuous or batch processing, and in either operation are a direct part of the production process. If on line, tanks may be filled or emptied often.

Off line. The mixing tanks are used as accumulators in which batches are placed for long-term storage. Filling and emptying are infrequent. The blending requirement is to produce a uniform batch from several different strata. This is known as stratified blending.

With on line blending, most of the mixing takes place in small, perfectly mixed zones of high shear rate near the impeller. Thus, overall blending time is related to the pumping capacity of the impeller and the frequency with which the tank contents pass through the impeller. Since tank contents are in motion and the components are added directly into the flowing stream, blend time is relatively short.

On the other hand, stratified blending has two steps:

- The blending step taking place in the zone of turbulence near the mixing turbine
- An erosion step, which is a function of the velocity at the interface between layers

The principles for top-entry and side-entry agitators are the same pertaining to stratified blending. Overall blending was found to be a function of the erosion of the interface between layers. Therefore, blend times are greatly extended because erosion takes place more slowly than does the actual blending. Mixing tanks that are operated intermittently suffer from having fluid momentum lost when the agitator is shut off. Thus mixing time is lost in building back up the momentum needed to generate high rates of erosion at strata interfaces. Qualitative statements about the two types of blending may be made as follows.

Stratified blending in storage tanks can result in undesirably long blend times. The mixer can be run during fillup to eliminate the problems associated with stratified blending. After the tank has been filled, the mixer can be shut down and restarted just before discharge for simple blending.

Adding ingredients at the surface can result in extended blend times by creating a situation resembling stratified blending.

Therefore, for stratified and on line blending, component additions should be made into the bulk flow created by the mixing impeller—below the surface and near the impeller, preferably in the direction of flow toward the impeller. To avoid creating excessive hydraulic loads on the impeller, and thus jeopardizing the mechanical integrity of the impeller or agitator, it is suggested that additions be made near the impeller, but not directed right at the blades. Blending is then controlled by the number of circulations made through the vessel, and takes place in the fastest manner.

The evaluation of time available is quite important, as it has a considerable affect on mixer horsepower. Input horsepower is selected to give a certain number of batch turnovers in a given time period. Mild blending applications typically require 1–3 batch turnovers per minute. Medium blending can be accomplished as a rule with 3–7 batch turnovers per minute. Vigorous blending requires on the order of 7–10 batch turnovers per minute. By extending the time period for blending, input horsepower can be decreased, or conversely, increasing the input horsepower will decrease the required blend time. The number of batch turnovers required to achieve a satisfactory blend is extremely variable. For example, 10–12 turnovers should give blending of readily miscible liquids of similar viscosity and density, such as alcohol and water. However, as many as three times that number of turnovers, i.e., 30–36 turnovers, may be required for readily miscible liquids of widely different viscosities and/or densities, such as glucose and water.

If a tank is filled slowly for a long period of time and then pumped out quickly when full, it is more cost effective to design an agitator than can blend the tank contents reasonable fast, and then run the agitator only just before emptying to produce uniformity. Continuous mixing in this case is not warranted.

C. BLEND TIME

The maximum time allowed for blending should be set by the requirements of the linked processes. For example, maximum blend time is 1 hour if the mixture from a tank must be sent downstream 1 hour after charging. However, blend time is often set arbitrarily short for convenience, at the sacrifice of higher capital and operating cost for the mixer.

Though we cannot relate power to blend time t directly for all agitation processes, we can use a rough empirical relationship:

$$P \propto t^{-a} \tag{1}$$

where: a = 1.5–2.5 for side-entry propellers
a = 2–3 for top-entry propellers and turbines
P = horsepower
t = blend time

Energy usage depends on both P and t, and can be estimated by:

$$\text{Energy} \propto t^{-a} \tag{2}$$

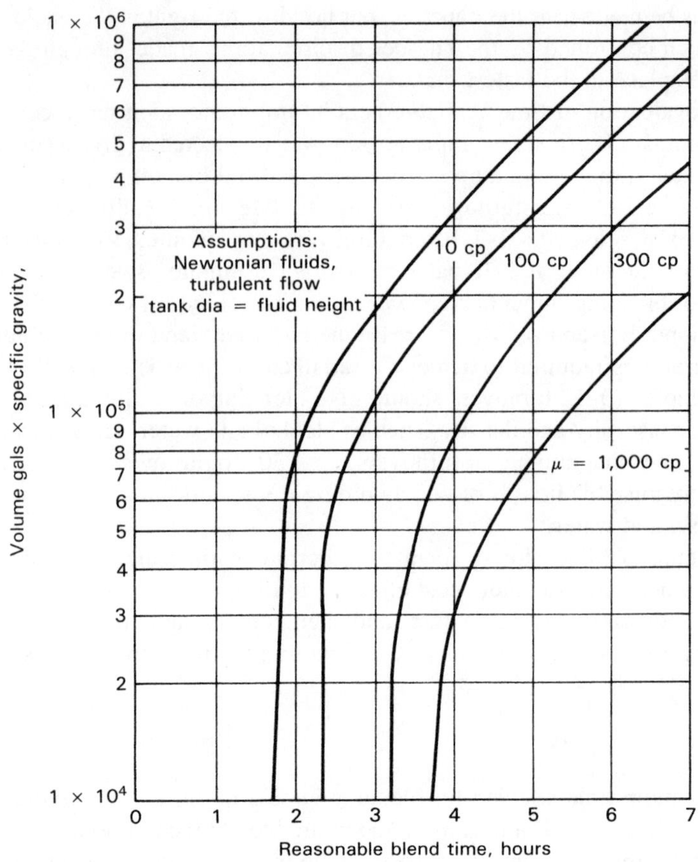

FIGURE 2-1. Approximate blend times recommended for side entry agitators.

Using Eq. (2) one can find the time related effect on horsepower to achieve blending. For example, we can find the effects of arbitrarily halving the allowed mixing time for a process:

- For a side-entry machine, power (and hence motor and drive) requirements would be 2.5–6 times what is really needed and energy usage would be 25–200% too high.
- For a top-entry mixer, power requirements would be 4–8 times what is really needed, and energy usage would be 50–300% too high.

Obviously, it costs dearly, in both capital and energy, to set blend time lower than need be. But increasing blend time does not necessarily save energy, because there is usually a minimum power level requirement to produce acceptable motion in the vessel. For example, at a low power level one may see good agitation around the impeller and minimum motion elsewhere. This may be perfectly acceptable for a paper-pulp high density storage chest application, where blending is not critical, except around the impeller and tank outlet, but not for a process that requires uniform and thorough blending of components.

How much time is reasonable for blending? It depends on fluid volume and viscosity, vessel geometry, and the degree of mixing desired. Extent is not often considered, but in fact the end-use of the fluid should govern the needed uniformity—not every fluid need be perfectly mixed. For example: a 20% variation in uniformity may be acceptable if a component is being diluted in one tank and then transferred to another for processing.

Figure 2-1 shows reasonable blend times for side-entry mixers in very large vessels, approximately 10,000 to 1,000,000 gallons in capacity, based on a few percent variation in fluid blending. These are empirical mid-range values. Experience has shown that by extending blend time up to twice these values reduces power. Beyond a twofold extension, the incremental power savings become insignificant.

For any blend time, it pays to take advantage of inexpensive means for getting more mixing before trying extra power. For example: retrofitting less efficient axial and radial flow turbines with more "flow efficient" hydrofoils, adding the components to be blended near the impeller instead of at the surface and running the agitator while components are being added to the vessel.

D. SAVING POWER IN BLENDING APPLICATIONS

It is not uncommon for an agitator to be operated at liquid levels lower than that designed for. For example, consider a mixer that blends a full load periodically but normally blends a partial load. Multiple side-entry agitators tied to level switches would be more operating cost effective (i.e., lower horsepower) than a single large machine, and the power saved could justify the added capital expense for the multiple agitators.

Variable speed mixers can also save power in applications where low speed is adequate for part of the blending time. Since power is proportional to speed cubed (for low viscosity applications), the savings are obvious. Potential applications include using high speed to blend components, then storing the blended final product in the same vessel with low speed agitation, perhaps periodic versus continuous operation. If the number of blend components is limited, and they are relatively similar in properties, i.e., specific gravity and viscosity of components and final product, a two-speed motor can provide cost effective speed variation. If the number of blend components is significant and/or if the specific gravities or viscosities of the components and final products are quite dissimilar, a variable speed drive can provide flexibility and operating cost savings.

An agitator may actually be turned off when blending is not required. For example, if a tank is filled slowly over a long period of time and then is pumped out quickly when full, why agitate continuously if the only purpose is to produce a uniformly blended product for discharge from the tank? It is significantly more cost effective to design and operate a mixer that can blend the tank contents reasonably fast, and therefore blend only before emptying.

E. HYDRAULIC OPTIMIZATION

The most direct avenue to power savings in flow controlled operations is hydraulic optimization. This is true optimization in that there are many ways of producing the required flow, and the objective is to choose the most cost efficient combination of mixing torque, turbine type, diameter, and speed.

As discussed in Chapter 1, all agitators deliver a combination of flow (pumping) and shear (head). Flow and shear directly consume power and it makes sense to reduce the factor that is not needed—i.e., minimize shear in a flow controlled application like blending.

The power required for turbulent flow (impeller Reynolds number above 10,000) can be expressed as

$$P = \rho Q H \tag{3}$$

$$P = N_p \rho N^3 D^5 / 1.53 \times 10^{13} \tag{4}$$

where: P = horsepower
ρ = specific gravity
Q = flow or pumping capacity $= N_Q N D^3$ (5)
H = shear or head development $= N^2 D^2$ (6)
N_p = power number for the turbine
N = turbine speed, rpm
D = turbine diameter, inches

The power number (N_P) is determined experimentally and is a function of the type of impeller, geometry and impeller Reynolds number (N_{Re}) where:

$$N_{Re} = D^2 N \rho / \mu \tag{7}$$

with μ = liquid viscosity.

Figure 2-2 shows values for several impeller types and geometries. In turbulent flow N_p is constant for a family of geometrically similar impellers. Therefore one can predict the effect of changes of impeller diameter or speed of rotation on power requirements. Agitator suppliers usually know values of

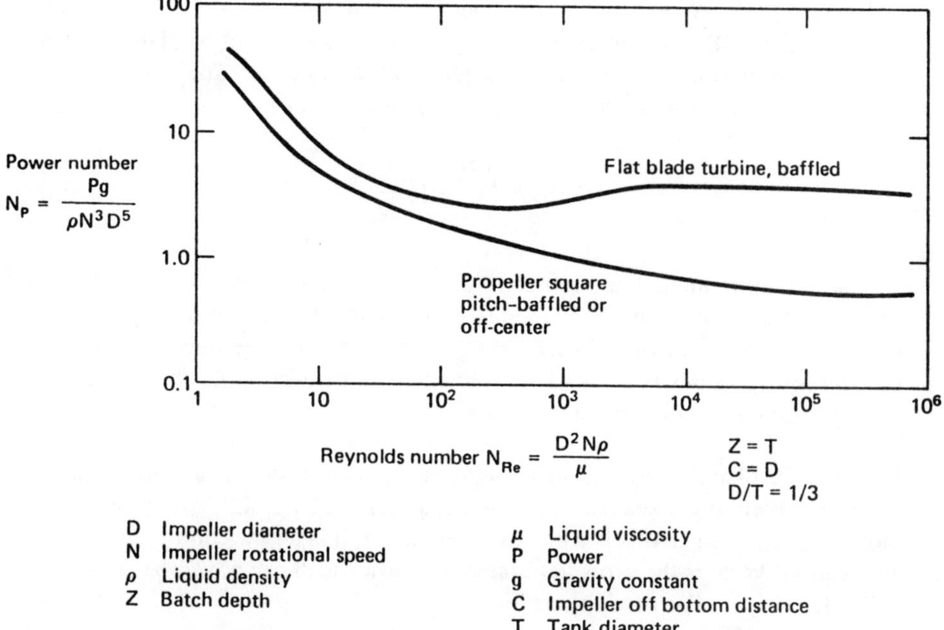

FIGURE 2-2. Power requirements for various impellers. Power number as a function of Reynolds number. (*Mixing Equipment, Inc.*)

N_p for various turbine types and geometries, and as a function of impeller Reynolds number, but they may consider them proprietary—especially the sophisticated hydrofoils. Power number can be calculated from Eq. (4) if power draw can be measured, and if the other variables are known (specific gravity, speed diameter of impeller).

The pumping number (N_Q), also known as the pumping efficiency, has seldom been determined, due to measuring difficulties. As with N_p, suppliers of agitators refrain from publishing values of N_Q, although they will often use it in a relative and comparative manner to evaluate alternative mixer selections. N_Q reaches a constant value in turbulent flow conditions, but the value here depends not only on impeller geometry but also on the ratio of impeller diameter to tank diameter. Side-entry mixing impellers are generally very small compared with vessel diameter so N_Q for turbulent flow is affected very little by changes in impeller diameter.

The parameter used for scaling flow-controlled operations, such as blending applications, is process torque: N^2D^5. Process torque has nothing to do with the torque rating of the agitator drive assembly. It is a measure of the work done on the fluid being mixed. Process torque is defined as delivered power divided by rotational speed, and is held constant in scaling flow-controlled agitation.

Knowing how N_p and N_Q behave, and how power is related to speed and diameter (Eq. (4)), we can evaluate how changing speed and/or diameter can deliver equal performance (process torque) with less power. This equation relates P, N_p, and N for agitation at constant process torque.

$$P_2 = P_1\left(\frac{N_{p2}}{N_{p1}}\right)\left(\frac{N_2}{N_1}\right) \tag{8}$$

Until the introduction and retrofitting of state-of-the-art hydrofoil designs as truly hydraulic optimum turbine configurations for flow-controlled processes, efforts at power reduction focused on geometrically similar impellers, where the ratio of N_p values is unity. The following example illustrates power reduction with geometrically similar turbines.

Example: A blending operation is being performed with a side-entry marine propeller. Blending takes 3½ hours and requires a 25 hp, 420 rpm (1800 rpm motor) agitator. Since this is a blending operation, it is a flow controlled mixing application. What is the propeller diameter? How much can power be cut while still achieving similar mixing results?

From Eq. (4),

$$P = N_p \rho N^3 D^5 / 1.53 \times 10^{13}$$

and assuming the power drawn by the propeller is 80% of full motor load, and

$$N_p = 0.87$$

$$\rho = 1.0 \text{ (waterlike liquid)}$$

$$25 \times 0.8 = (0.87)(1.0)(420^3)D^5/1.53 \times 10^{13}$$

rearranging for D:

$$D^5 = 4.7474 \times 10^6$$

therefore

$$D = 21.6 \text{ inches}$$

Now let us look at the option of using a lower-speed motor with the same drive speed reduction ratio. (We could also evaluate changing the reduction ratio by changing gears or the drive itself). With a 1200 rpm motor, the agitator speed will be reduced to 280 rpm. Since N^2D^5 must remain constant to achieve similar mixing results, impeller diameter would have to increase by approximately 18% (i.e., D increases to 25.5 inches). The effect on power is predicted by Eq. (8):

$$P = 25(230/420) = 16.7 \text{ hp}$$

This mixer would require a 20 hp motor in practice, resulting in at least a 5 hp savings while achieving the same mixing result. If instead we reduced the motor speed down further to the next lower commercially available speed of 900 rpm, the agitator speed would reduce to 210 rpm, and power would be only 12.5 hp for constant process torque. Such a design would use a 15 hp motor, resulting in a net power savings of 40% over the original 25 hp unit. Further speed reductions would save more power, but would not be practical for the following reasons:

- Lower speed motors (below 900 rpm) are not standard and commercially available, and as such become special order items that are very expensive.
- Greater speed reduction off the original motor could be very inefficient, given the type of speed reduction (i.e., belt drive, gear box, etc.). Change gears may not be available to accommodate speed reductions in gear box drives.
- Impeller diameter would become so large that an oversized drive and shaft assembly would be needed to support the extra weight and the tank walls would restrict flow development.

The typical measure of mixing intensity in a blending operation of two or more miscible liquids is blend time, that is, how long it takes after the last component is added to a tank for the mixture to become homogeneous. The blend time of a given system will be a function of the degree of agitation within the mixing vessel, and the type of application or optimal process requirement will dictate what degree of agitation and blend time are required. Most typical applications can be labeled with an intensity of blending requirement along with a blend time. A few examples are shown in Table 2-1.

Blend time may not always be the only measure by which a mixer is selected for a blending operation. Considering it from a different viewpoint, a certain process may require a certain degree of blending rather than a completely uniform mix. For instance, in a storage operation where product is being drawn off from the bottom of a tank near the impellers, homogeneity throughout the entire tank would probably not be a critical factor. In contrast, feed for a chemical reactor may be drawn from the top of a tank in a process where the reaction is very sensitive to changes in feed concentration. This would necessitate a completely uniform and homogeneous blend.

TABLE 2-1. Typical Blending Requirements.

Typical Application	Intensity of Blending	Blend Time	Relative Fluid Velocities
Storage	Mild	30 min	1.0
Neutralization			
Fuel additive blending			
Batch formulation	Medium	12 min	2.5
pH control			
Heat transfer			
Continuous	Violent	6 min	5.0
Flash mixing			
Pigments blending			

Mixing time studies in the jet mixing of a tank have demonstrated that mixing time θ is related to the flow rate Q and the head H of the jet stream:

$$\theta = \frac{0.017T^2}{Q^{0.5}H^{0.25}} \tag{9}$$

with T being tank diameter.

In a tank being mixed by an impeller, the head developed is analogous to the square of the jet velocity and hence mixing time is a function of the product of flow Q and impeller velocity V, i.e., a function of QV.

Mixing time in a tank is a function of momentum flux or QV as indicated below:

$$\theta = \frac{9.1V^{0.5}}{QV^{0.41}}\left(\frac{\mu}{P}\right)^{1/6} \tag{10}$$

where μ is viscosity and P is power.

Thus momentum flux or QV of an impeller has been used to successfully correlate mixing or process results.

The rationale behind the momentum flux concept can be explained as follows: An impeller in a tank emits a jet stream. The stream has a certain velocity, flow, and cross-sectional area. As the jet expands further away from the mixing impeller within the confines of the tank, the area, flow, and velocity change markedly as additional fluid is entrained. However, momentum or QV is conserved except for the retarding forces from shear at the tank walls. Knowing the momentum of the stream leaving the impeller, the momentum in the tank (i.e., in the expanded jet) can be determined. Thus the momentum or QV of the impeller which manifests itself as velocity of motion in the tank, is an accurate measure of tank motion, and the mixing or process result. Mixing times and tank velocities can be accurately predicted by considering the QV at the impeller, and therefore momentum flux can be a basis of mixer selection.

There are numerous relationships that predict QV or momentum output from an agitator. The momentum can be related to the impeller dimensions and speed by equations similar to the power equations, with different exponents, i.e., Eqs. (9), (10). Another relationship conveniently relates QV to the horsepower and speed of an impeller, and the constant in this relationship is a measure of the mixing efficiency of an impeller. These relationships are shown in the following formulas:

$$QV \propto C(ND^3)(ND) \propto CN^2D^4 \quad (11)$$

$$QV = \left(\frac{C}{N_p^{0.8}}\right)\frac{H_p^{0.8}}{N^{0.4}} \quad (12)$$

where C is an impeller constant similar to a drag coefficient
N_p is impeller power number
N is impeller speed
D is impeller diameter
H_p is horsepower

The value of $(C/N_p^{0.8})$ in Eq. (12) is a measure of efficiency of a particular impeller type and geometry. It is constant for a specific type and geometry turbine. As one can imagine, mixer suppliers keep the values of C and N_p confidential.

QV can be used to select agitators for specific results. The tank velocity or degree of agitation can be predicted by considering the QV to tank volume ratio as given by:

$$\text{Process Result} = QV/\text{Vol}^{2/3} \quad (13)$$

Thus $(QV/\text{Vol}^{2/3})$ can be easily used as a scaleup quantity from existing or experimental installation.

The average velocity in a tank is not Q/A, with A being cross-sectional area. Velocity varies with the square root of QV. It is the average tank velocity that determines the degree of mixing to be expected. Also, it is the basic requirement for scaleup to establish equal process result.

One can move from QV to mixer selection by determining the basic QV required by considering tank volume and degree of agitation required. (i.e., Eq. (13)). This basic QV is modified by factors to account for fluid properties and tank shape to give the process QV. Then the desired impeller power and speed is chosen (by Eq. (12)) to provide the required process QV.

F. SOLIDS SUSPENSION

Suspension of solids is probably the most common application of mixing technology. In order for solids to be suspended, more important than just flow, is the flow per unit area, or velocity that the mixer imparts to the particles. This velocity must be greater than the terminal settling velocity of the solids or the particles will settle to the bottom of the tank.

Axial flow impellers and the flow pattern they produce are best suited for suspending solids. The optimum location of a single or lower turbine is 1/3 impeller diameter off the bottom of the tank. Changing this distance increases the power required for solids suspension.

The variables that are important in a solids suspension mixing application are:

1. How difficult is it to suspend the solids? This is determined by the density and viscosity of the liquid, and the density, mesh size, and percentage of solids. These factors are used to calculate the settling velocity of the particles in the particular liquid. As the particles become larger and or more dense, the settling velocity increases. As the liquid becomes more viscous and/or more dense, the settling velocity of the solids decreases.

It makes no difference whether the liquid moves past the particle or if the particle moves through the liquid. A particle falling under the action of gravity will accelerate until drag force just balances gravitational force, after which it will continue to fall at a constant velocity known as the terminal or free-settling velocity U_t, given by:

$$U_t = \sqrt{\frac{2gM_p(\rho_p - \rho)}{\rho\rho_p A_p C}} \text{ ft/sec} \qquad (14)$$

where: g = local acceleration due to gravity, 32.17 ft/sec
M_p = mass of the particle, lb
ρ_p = density of the particle, lb/ft^3
ρ = density of the liquid, lb/ft^3
A_p = projected area of the particle in the direction of motion, ft^2
C = drag coefficient, dimensionless.

The drag coefficient C has been found to be a function of the shape of the particle and the Reynolds number, $D_p \rho u_t/\mu$,

where: D_p = diameter of particle, ft
μ = liquid viscosity, lb/ft-sec

For the case of spherical particles, Eq. (14) becomes:

$$u_t = \sqrt{\frac{4gD_p(\rho_p - \rho)}{3\rho C}} \text{ ft/sec} \tag{15}$$

For Reynolds number $N_{Re} < 0.3$,

$$C = 24/N_{Re} \tag{16}$$

This corresponds to Stoke's law. The terminal settling velocity in the Stokes' law region (i.e., where $N_{Re} < 0.3$) becomes:

$$U_T = \frac{gD_p^2(\rho_p - \rho)}{18\mu} \text{ ft/sec} \tag{17}$$

In the Newton's-law region, which covers the range of $1{,}000 \leq N_{Re} \leq 200{,}000$, drag coefficient C has an approximate value of 0.44 for spheres. In this region, Eq. (14) becomes:

$$U_t = 1.74\sqrt{gD_p(\rho_p - \rho)/\rho} \text{ ft/sec} \tag{18}$$

In the intermediate region, where $0.3 \leq N_{Re} \leq 1{,}000$, the drag-coefficient relationship for spheres can be approximated by:

$$C = \frac{18.5}{N_{Re}^{0.6}} \tag{19}$$

This value of drag coefficient can be used with Eq. (15) to find the terminal settling velocity of the particle.

As shown above, the terminal settling velocity of a spherical solid is dependent on which region, and hence which formula, is applicable. The particle Reynolds number must be calculated *after* assuming a particular region is applicable and calculating the terminal settling velocity based on that region's formula, i.e., Eq. (17) or (18). The particle Reynolds number must fit the particular limit or range of the region's formula for the calculated value of U_t for that region to be valid. If the Reynolds number check indicates that another region's formula for U_t should be used, then terminal settling velocity should be recalculated based on the correct region and formula.

Most particles in solid suspension mixing applications are not spherical in geometry, but are irregularly shaped. The drag on a nonspherical particle depends upon its shape and its orientation with respect to the direction of motion.

For the same drag coefficient, spherical particles exhibit greater Reynolds numbers and hence greater settling velocities than irregularly shaped particles, disk shaped particles, and cylindrically shaped solids. Therefore calculating terminal settling velocity based on spherical particles will provide some conservation in the calculation if the true particle shape is nonspherical. The mixer will be designed for particles that settle slightly faster than those in the application.

As a general rule of thumb, the rise velocity of the liquid-solid slurry created by the mixer (calculated as impeller pumping capacity divided by annular area between the tank wall and impeller) should be 5–10 times the settling velocity of the largest or most dense particle, to achieve suspension off the bottom of the tank. For uniform suspension of all particles throughout the mixing tank, the rise velocity should equal 7–10 times the settling velocity.

Particles with a settling velocity greater than 1 ft/minute are called free settling solids. These solids settle at their natural terminal velocity (up to 20 ft/min). Particles with a settling velocity less than 1 ft/minute are called hindered settling solids. When the concentration of solids becomes high, the particles will be close enough to each other to exert a mutual retarding effect on settling. This condition creates hindered settling of the particles.

Hindered settling particles settle so slowly that they typically are treated not as solids suspension requirements but as blending or fluid motion applications. Examples of hindered settling slurries include biological-floc suspensions, paper stock suspensions, and fine grind minerals processing slurries.

Figure 2-3 shows the effect that concentration of solids and specific gravity driving force (i.e., difference between solid and liquid specific gravity) has on determining whether the solids will be free settling or hindered settling.

Another factor determining whether a solid will exhibit free or hindered settling characteristics is the mesh size and particle size distribution. Generally speaking, the finer the particle size and/or the narrower the particle size distribution range, the more likely the solids will demonstrate hindered settling. Typically solids finer than 200 mesh in size, in concentrations greater than 40% by weight, will exhibit hindered settling.

2. *Degree of suspension required.* Three categories of solid suspension are traditionally defined. They are:

(a) On-bottom motion, in which heavy and/or coarse particles are moving on the tank bottom, without any progressive fillet formation (i.e., without a continuous buildup of solids along the tank bottom, typically in the corners where tank bottom and wall come together). Lighter and/or finer solids will be suspended off the bottom of the tank.

(b) Off-bottom suspension, in which all particles are suspended off the tank bottom, but not uniformly throughout the tank contents. Some particles, particularly those smaller in size and/or lighter in gravity will be suspended more uniformly than others.

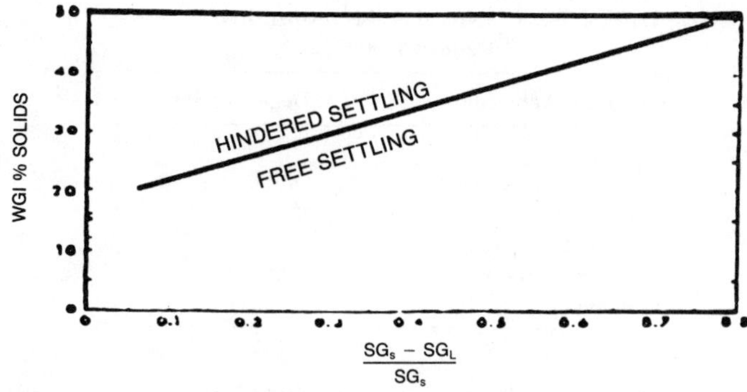

FIGURE 2-3. Free versus hindered settling. (*Mixing Equipment, Inc.*)

(c) Complete uniformity or uniform suspension, in which all particles, large and small, heavy and light, are suspended uniformly throughout the tank contents.

Care should be taken in specifying the appropriate degree of suspension required. Complete uniformity requires more mixer power than off-bottom suspension or on-bottom motion. The relative power, speed, and torque ratio needed to achieve each type of suspension is shown in Table 2-2.

Note not only a premium on power, but also on torque (and therefore higher capital cost) to achieve complete uniformity of suspension over off-bottom suspension and/or on-bottom motion.

As illustrated, specifying greater suspension than is necessary for your process, wastes not only power, but also at the expense of higher capital cost. In most cases, off-bottom suspension is adequate for most solid suspension applications.

The degree of suspension, like blend time, will be dictated by the nature of the application and actual process requirements—point of drawoff, dissolving time, etc. Typical solid suspension applications and their required degree of suspensions are listed in Table 2-3.

3. Tank size and shape. This is much more important in solids suspension than in blending. As the batch depth to tank diameter ratio increases, tanks

TABLE 2-2. Relative Power and Torque for Different Degrees of Suspension.

Criteria	Hp Ratio	Speed Ratio	Torque Ratio
Complete uniformity	25.0	2.9	8.6
Off-bottom suspension	5.0	1.7	3.0
On-bottom motion	1.0	1.0	1.0

TABLE 2-3. Typical Solids Suspension Requirements.

Typical Application	Degree of Suspension
Surge tank	
Slurry storage	Off-bottom
Dissolving	
Continuous reaction	
Overflow slurry drawoff	Uniform
Leaching	

become taller and the solids must be suspended further up requiring more horsepower and multiple turbines. At batch depth to tank diameter ratios exceeding 1.25 (i.e., $Z/T > 1.25$), it is recommended that dual turbines be used for solids suspension applications. See Figs. 2-4 and 2-5. Dish bottom tanks of 2 : 1 ellipsoidal shape are ideal for solids suspension as the smooth contour and transition from bottom to side wall aid in directing the flow along the bottom and up the sides of the tank. Steep cone bottom tanks are poor shapes for solid suspension applications because solids tend to "cyclone out" in the heel of the cone and can be very difficult to remove or suspend—often plugging outlets traditionally located in the cone.

In solids suspension, not only the economics of the agitator are involved, but the total tank/agitator package must be considered as a system together.

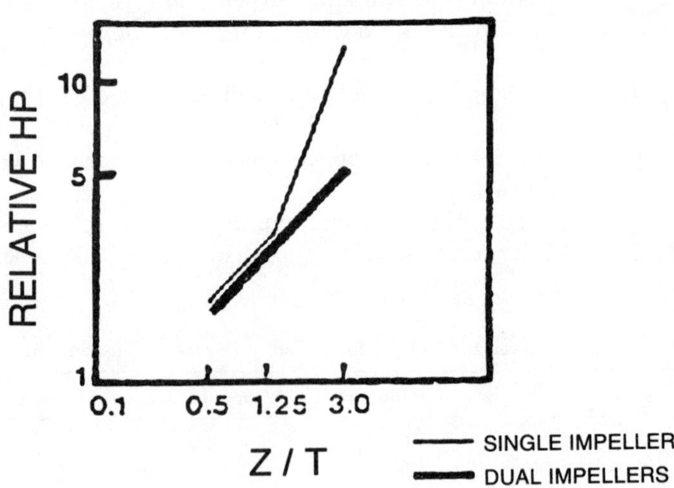

FIGURE 2-4. Relative Hp versus Z/T. (*Mixing Equipment, Inc.*)

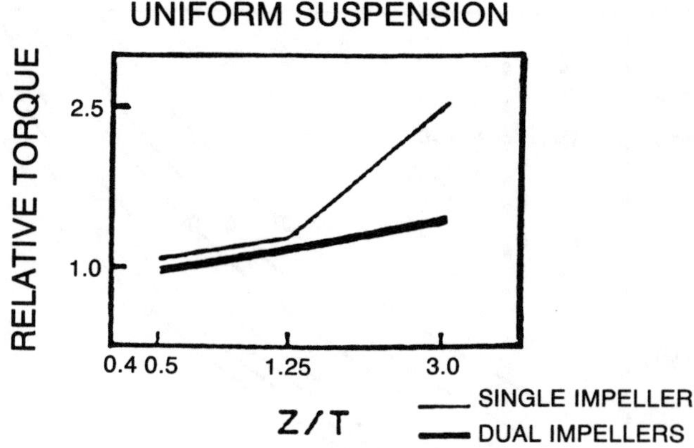

FIGURE 2-5. Relative torque versus Z/T. (*Mixing Equipment, Inc.*)

Each process must be evaluated individually to determine the optimum configuration and agitator.

Once again, the main points to consider in solids suspension mixing applications are:

1. Settling velocity of the solids
2. Degree of suspension required
3. Optimum tank shape

As previously discussed, a key variable to consider when designing a mixer for a solids suspension application is the terminal settling velocity of the particles being suspended. Settling velocity is a function of particle geometry, size, and density. The settling velocity can be calculated based on particle Reynolds number, and an appropriate formula based on whether Stokes', Newton's, or an intermediate law is applicable.

Alternatively, the settling velocity can be determined experimentally by measuring the time it takes solids to drop a known distance in the liquid in question. This requires that the liquid be clear in order to see the falling particles. Shape correction factors can be determined experimentally by comparing the settling velocity of an actual particle to that calculated for a spherical particle of similar density and size via one of the formulas (Stokes', Newton's, or intermediate law).

Another way to estimate the settling velocity of particles is by using the chart in Fig. 2-6.

There are many agitator designs and selections to achieve the desired degree of solids suspension. Several criteria can be used to make relative judgments of

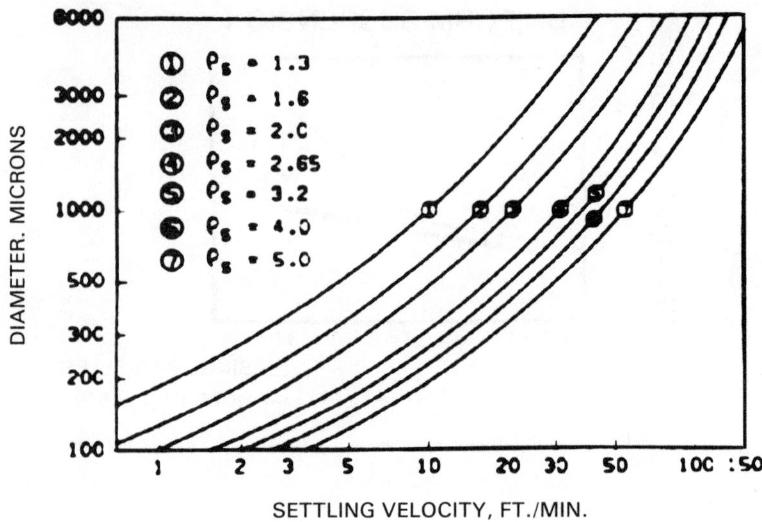

FIGURE 2-6. Free settling rate in water versus particle size. (*Mixing Equipment, Inc.*)

value between different agitator selections. For each process requirement, torque, power, pumping capacity and shear rate will have different relative values. For any specific solid suspension application the efficient use of flow, torque, or power will be affected by other considerations such as impeller type, location and impeller diameter to tank diameter ratio, D/T.

A wide range of mixer selections can be used depending on the process requirement. This can be shown in Fig. 2-7. The interrelationship of process requirement, agitator fluid mechanics and vessel geometry and design must be recognized when addressing a solid suspension mixing requirement. Case sizes 1, 2, etc. represent various gear box speed reducer sizes determined by torque limitations (i.e., case size 2 has greater torque limit than case 1, etc.).

Relative torque is a means of comparing mixer selections to achieve desired levels of solid suspension. In addition, the efficiency of using the torque (i.e., drive efficiency for a given combination of power and speed reduction) must also be considered. For each process, the torque represented by a specific drive size must be applied differently to achieve optimum results (i.e., different combinations of horsepower and speed N).

Figure 2-8 shows that for constant process results in a solids suspension application, where the solids exhibit free settling characteristics, torque is minimized at an impeller to tank diameter ratio of approximately 0.25. This is the optimum design point with regard to torque, and since torque is a direct reflection on capital cost, this selection would represent the most capital cost effective mixer to achieve the desired degree of suspension.

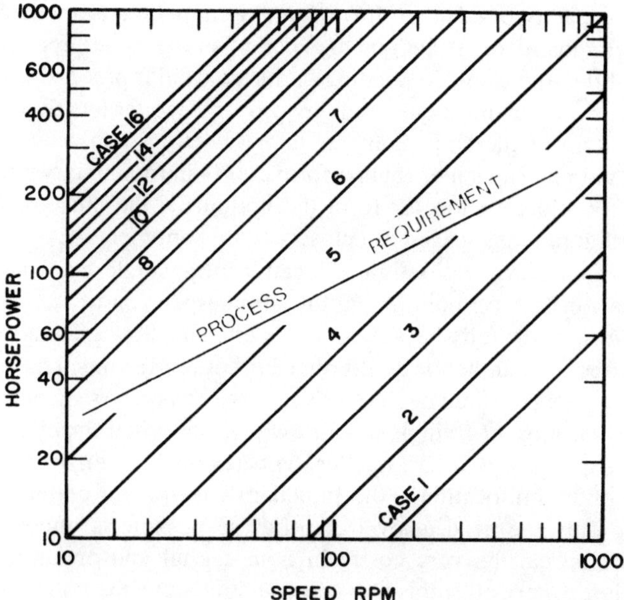

FIGURE 2-7. Various mixer selections for equal process results. (*Mixing Equipment, Inc.*)

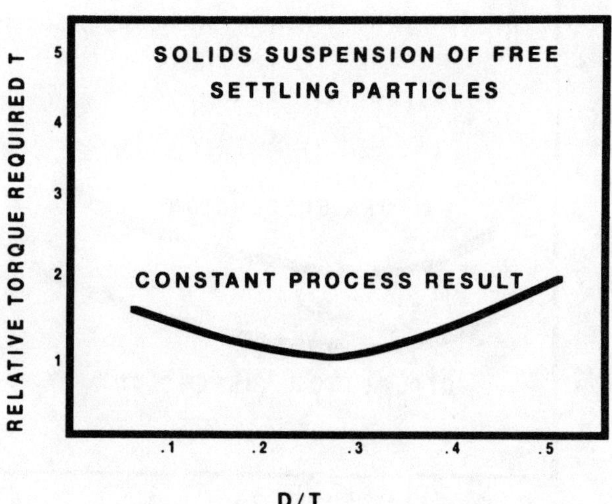

FIGURE 2-8. Effect of D/T on torque to achieve equal process results. (*Mixing Equipment, Inc.*)

Pumping capacity or primary flow of mixing impellers is another method of comparing various mixer selections to accomplish the same process result pertaining to solid suspension. Depending on the particular process result, the calculated flow is utilized most efficiently by proper impeller location with respect to off the bottom of the tank. Some agitator selections can achieve very high theoretical pumping capacities, but perform poorly in mixing tanks because they are located too close or too far from the bottom of the tank, or in multiple impeller configurations, spaced too closely to one another.

Figure 2-9 indicates the optimum location of a single axial flow constant pitch turbine for both off-bottom and uniform suspension of free settling slurries. Axial flow hydrofoils because of their efficient flow generation and pure axial flow direction can be located further off bottom (as much as one impeller diameter) to achieve desired degrees of solid suspension. As shown in Fig. 2-9, the optimum location of a single axial flow constant pitch impeller for off-bottom solids suspension is 0.33 impeller diameter off bottom. For uniform suspension the optimum location is 0.5 impeller diameters off bottom.

Choosing only the needed level of solids suspension is important because overspecification can be very costly in both capital and operating cost. On-bottom motion, where all solids move but seldom leave the bottom of the tank, is enough mixing if the only purpose is to keep the solids from compacting.

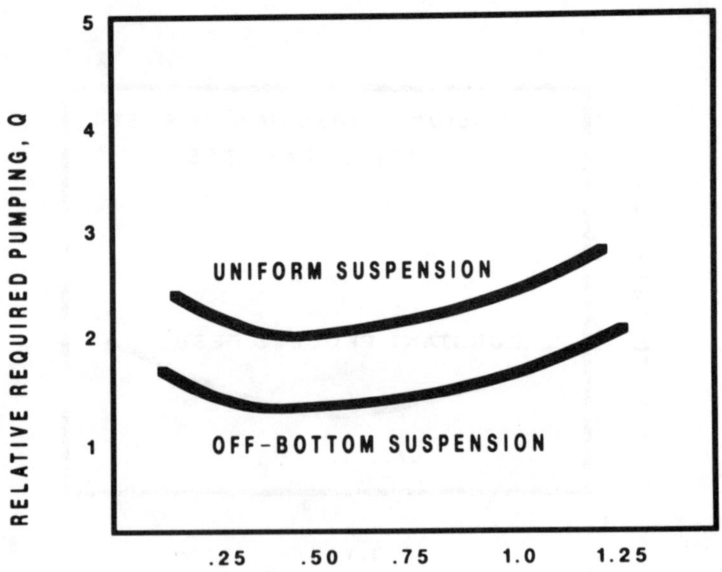

FIGURE 2-9. Effect of impeller location on pumping and degree of solid suspension. (*Mixing Equipment, Inc.*)

For example, basic sedimentation in oil storage tanks requires on-bottom movement to keep it from compacting and progressively building up. To compare power requirements for different levels of solids suspension, let us assign a relative power requirement of 1 for on-bottom motion.

Off-bottom suspension has a relative power requirement of 3–5 in low viscosity slurries with free settling particles. The factor 3 reflects suspension of the largest and/or heaviest solid to a depth of approximately 30% of the batch. The factor 5 represents suspension of the heaviest or largest solid to a depth of approximately 50% of the batch. This level of solids suspension will keep solids off the tank bottom most of the time, unless there is an upset in the process and a larger particle size distribution is introduced into the mixing tank.

Off-bottom suspension is appropriate when trying to dissolve solids because it ensures adequate circulation. The key to solids dissolution is to achieve off-bottom suspension of solids and provide adequate flow velocity around the particles as a driving force for dissolving.

Uniform suspension typically means suspending solids so that the concentration is nearly equal at all points in the vessel up to about 95% of the batch depth. Applications include solids removal by surface overflow, catalytic reactions such as hydrogenations where the absence of solids in part of the batch could produce unwanted side-reactions and/or hot spots. The relative power requirement for uniform suspension is about 25 times that for on-bottom motion.

Table 2-4 shows how power and settling velocities vary for classification of solids suspension.

At about 40% by weight solids, due to the increase in apparent viscosity, the slurry behaves as if it were a non-Newtonian fluid. The power requirement drops at this point but then increases as the solids concentration is further increased (see Fig. 2-10).

There is a maximum solids concentration for a given mixing system, beyond which the fluid motion stops and any reasonable increase in applied power does not restore motion. This maximum concentration is called the ultimate weight

TABLE 2-4. Power Requirements for Various Degrees of Suspension and Settling Velocities.

	Power at a Settling Velocity of:		
Degree of Suspension	16–60 ft/min	4–8 ft/min	0.1–0.6 ft/min
Complete uniformity	25	9	2
Off-bottom suspension	5	3	2
On-bottom motion	1	1	1

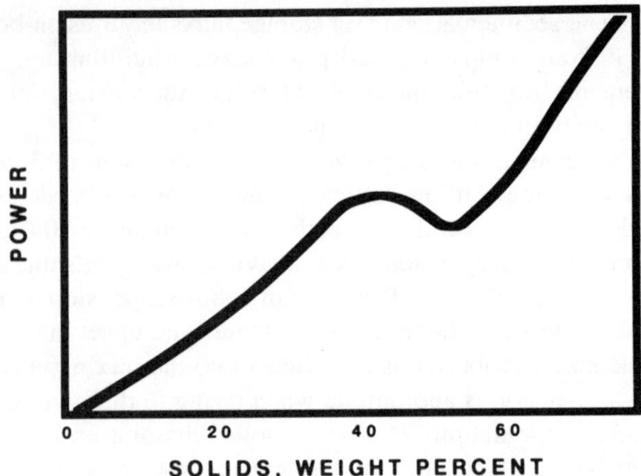

FIGURE 2-10. Power requirements for constant degree of suspension versus solids content. (*Mixing Equipment, Inc.*)

percent settled solids. It reflects the percentage by weight the solids have when they are allowed to settle in a tank, drawing off the supernatant liquid collected above the bed of solids. In other words, it is the highest solids concentration allowing for liquid to surround the particles in the settled bed. Typically, ultimate weight percent settled solids will range from 70 to 90%.

Mixing at this high concentration of solids has the advantage that coarse particles will not settle out. However, the major disadvantage is that the mixer must operate in a highly viscous, non-Newtonian slurry application. The viscosity of the slurry will depend on particle size distribution, as well as solids concentration. Finer solids fractions (above 250 mesh) will increase slurry viscosity over more coarse fractions. A fine grid slurry takes less power at lower percent solids and more power at higher percent solids than does a coarse grind.

In consideration of the basic design of an agitator for a slurry storage tank, one must take into account the behavior of the specific slurry being processed. Very often the particle size distribution of the solids will be set based on upstream or downstream process requirements or economics. For example, a coal slurry being stored at both ends of a pipeline (i.e., the mine side and the power plant side) will have the coal particle size distribution established to optimize the cost of transporting the slurry through the pipeline. This distribution may therefore be far from optimum with regards to the agitator selection for the slurry storage tanks.

For a given particle size distribution, there is an optimum solids concentration based on process requirements and overall cost. The slurry solids concentration can greatly affect slurry apparent viscosity. Fig. 2-11 shows the effect that solids content has on viscosity in a coal slurry.

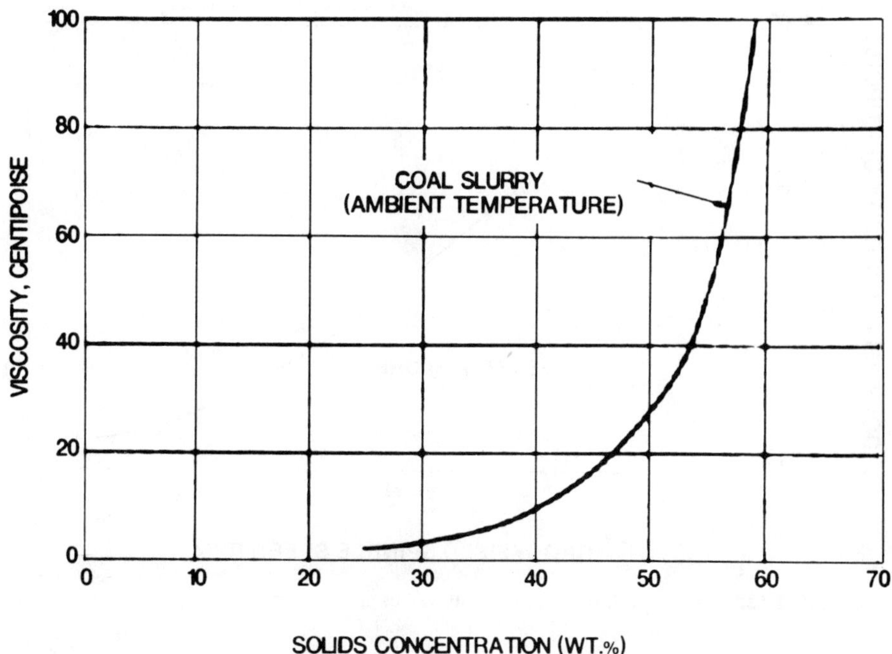

FIGURE 2-11. Effect of solids concentration on slurry viscosity. (*Mixing Equipment, Inc.*)

At high concentrations, slurries are usually non-Newtonian in nature. Their apparent viscosities decrease with increasing shear rates. In other words slurries at high solid concentration tend to be "shear-thinning" at higher mixer speeds. Shear rate varies within an agitated tank depending upon proximity to the impeller.

In the case of a non-Newtonian slurry, it is desirable to measure the viscosity with a mixing type viscosimeter in which shear rates are comparable to the average values that will be encountered in the full scale plant. Paddle type viscosimeters have been used to measure slurry viscosity but Brookfield viscosimeters are most commonly used. The Brookfield has spindles that operate at controlled and known speeds immersed in the slurry. Accurate measure of slurry viscosity as a function of spindle speed is obtained with Brookfields as long as solids do not collect on top of the spindle itself. Fig. 2-12 depicts Brookfield viscosity versus spindle speed of a typical coal slurry at 55% by weight solids. It should be readily apparent that a good knowledge of average impeller shear rates and average shear rates in the bulk contents of a tank is necessary to properly design the mixing system. Remember, the impeller "sees" only the apparent viscosity within its blade periphery based on the impeller shear rate, and power draw will be predicated on apparent viscosity "seen" by the turbine.

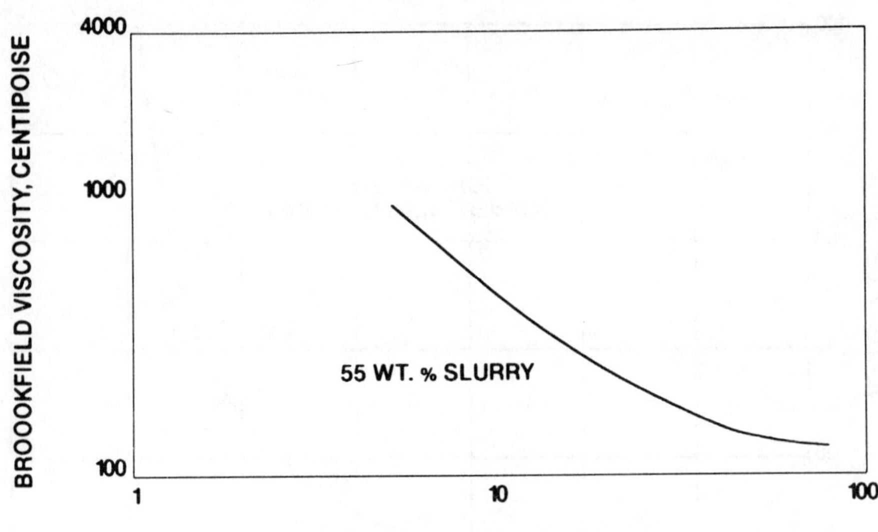

FIGURE 2-12. Apparent Brookfield viscosity versus spindle speed. (*Mixing Equipment, Inc.*)

Horsepower drawn by an agitator in a slurry storage and solids suspension application is dependent upon solids concentration. Fig. 2-13 shows the dramatic increase of power requirement with increasing solids content, based on the ratio of percentage of solids to the ultimate settled solids percentage. Ultimate settled solids percentage is affected by particle size distribution, density of particle and geometry of the solids. For example, let us consider a slurry at 55% by weight solids, if the percentage by weight solids in the ultimate settled material is 70%:

$$\% \, U = \tfrac{55}{70} = 79\%$$

If another slurry at 55% by weight solids is such that the percent by weight solids in the ultimate settled material is only 60%, then

$$\% \, U = \tfrac{55}{60} = 92\%$$

The particular ratio U is much more important in determining process requirements than is the actual percentage solids itself. The main criterion is how close one is to the ultimate settled solids percentage of the materiel being suspended. This type of approach is normally used above 40% of the ultimate settled solids percentage.

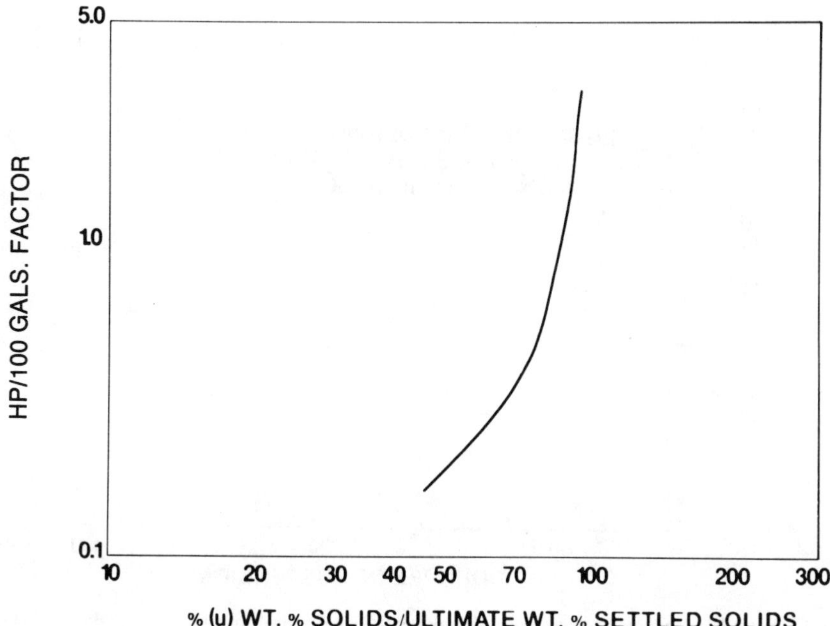

FIGURE 2-13. Effect of weight percentage of solids (ratio to final settled solids, % u). (*Mixing Equipment, Inc.*)

Degree of suspension is defined as the ratio of weight percentage of solids in suspension to the weight percentage of solids in the mixing vessel. As can be seen in Fig. 2-14, the impact of degree of suspension on power requirements is rather significant, particularly at high degrees of suspension uniformity. Conventional "uniform" suspensions occurs at 95% degree of suspension. To achieve greater degrees of suspension requires inordinate relative power levels. For example, to go from 98 to 99% degree of suspension requires 2½ times more power!

Power savings may be realized by allowing some of the solids to settle in the corners of the tank where the bottom transitions to the tank wall. This is especially true in large diameter tanks where suspension of solids without corner settlement can be prohibitively expensive regarding capital and operating cost of the agitator(s). Corner settling of solids is known as formation of a fillet. Fig. 2-15 depicts corner filleting. Fillets of 5–10% of the tank height occupy only 1–2% of the tank volume. This and other relationships between fillet size and percentage of tank volume (for the case of liquid height Z equal to tank diameter, $Z/T = 1$) is shown in Fig. 2-16.

Figure 2-17 shows relative power savings realized by using larger fillets. Fillets can be either natural, i.e., formed by the solids, or manmade and built in during tank fabrication.

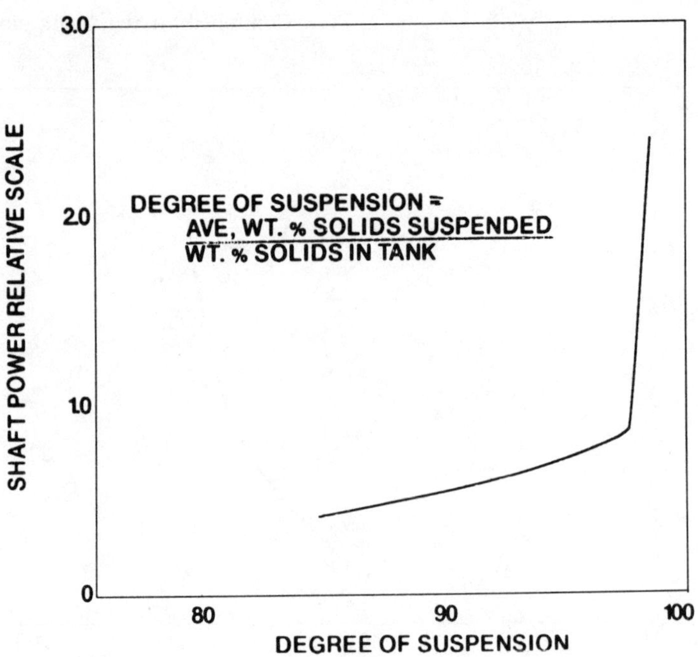

FIGURE 2-14. Degree of suspension versus relative shaft power at a typical weight percentage slurry in the system. (*Mixing Equipment, Inc.*)

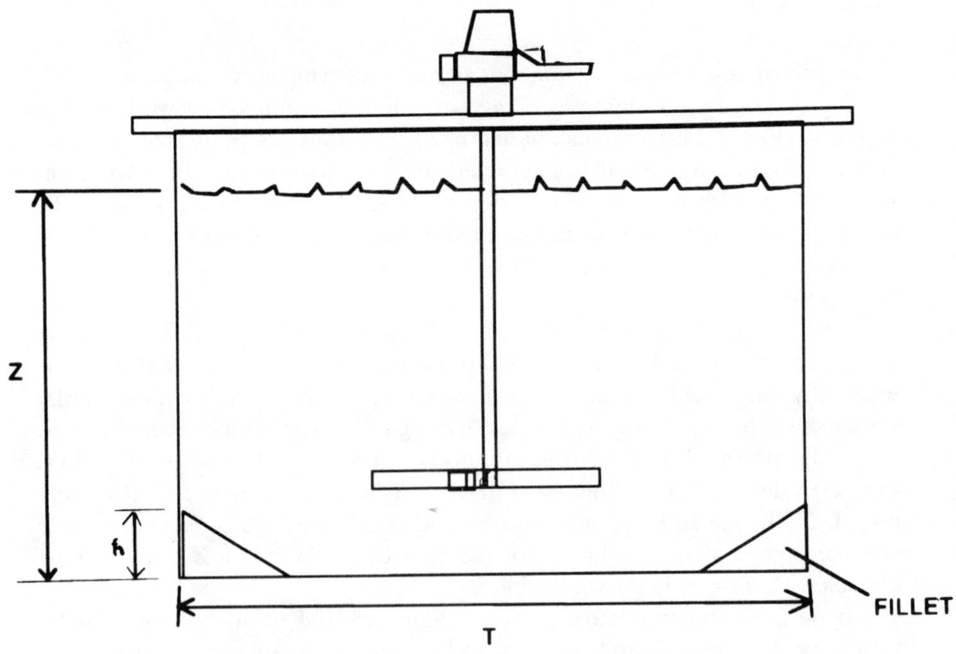

FIGURE 2-15. Fillet size. (*Mixing Equipment, Inc.*)

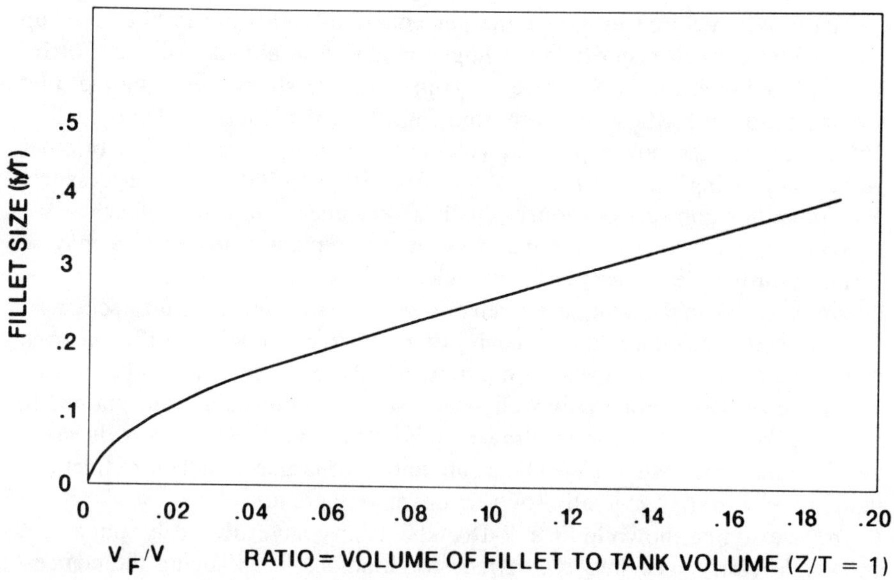

FIGURE 2-16. Fillet size versus ratio of volume of fillet to tank volume. (*Mixing Equipment, Inc.*)

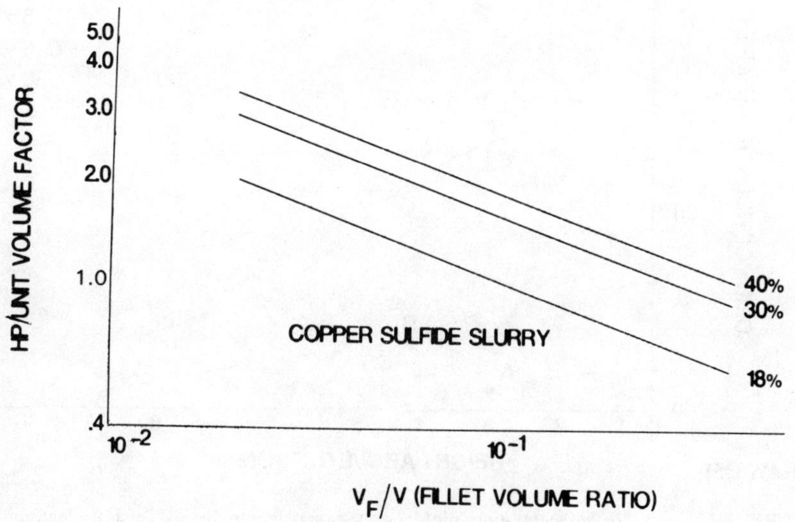

FIGURE 2-17. V_F/V (fillet volume ratio). (*Mixing Equipment, Inc.*)

The power required to reduce the percentage of solids in the fillet goes up incrementally much faster than the improvement in total tank volume. For instance, looking at the 30% by weight copper sulfide slurry, reducing the fillet volume ratio from 8% to 4% (or from Fig. 2-16 the fillet size from 23% to 16%) increases the power per unit volume factor from 1.75 to 2.25. In other words, improving the "active" volume from 92% to 96% would necessitate 30% more horsepower! Obviously this is a very uneconomical approach.

With a mixture of particle grind sizes present in many slurries, there may be variation in the sieve analysis of particles (i.e., particle size distribution), at different heights in the tank, at a given power level. In continuous flow schemes, this means the residence time of each grind size fraction will be different from that of the others. The elevation of a drawoff tube or "riser" outlet becomes a key factor in determining how well steady state is achieved and maintained in a mixing vessel. An incorrect drawoff tube elevation or an improperly sized overflow tube can lead to a steady accumulation of larger particle size fractions in the tank bottom, eventually shutting down the agitator.

In the example shown in Figs. 2-18 and 2-19, we have arbitrarily split a 30% by weight slurry feed into four grind size fractions, "A" being the coarsest particles and "D" the finest. The only process requirement is that the outlet composition be the same as the inlet feed material. Depending upon the location

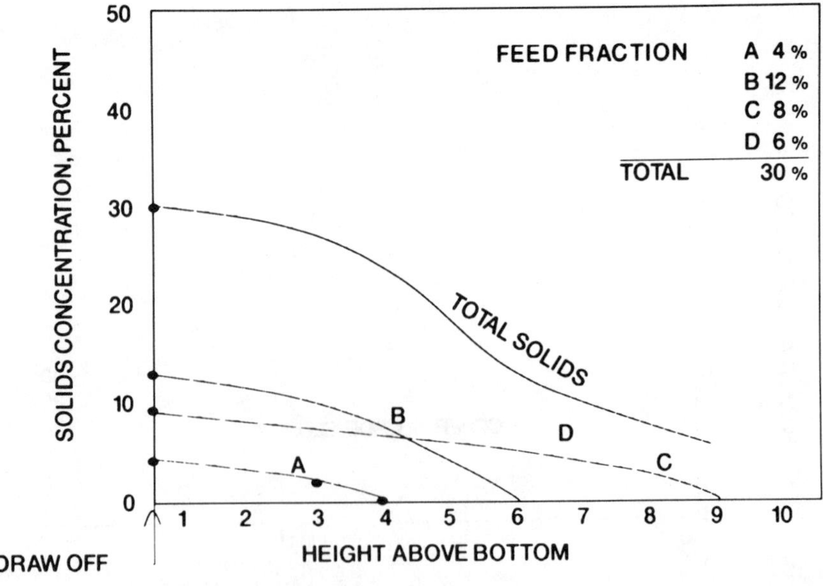

FIGURE 2-18. Solids concentration profile at various depths for 30% slurry and bottom drawoff. (*Mixing Equipment, Inc.*)

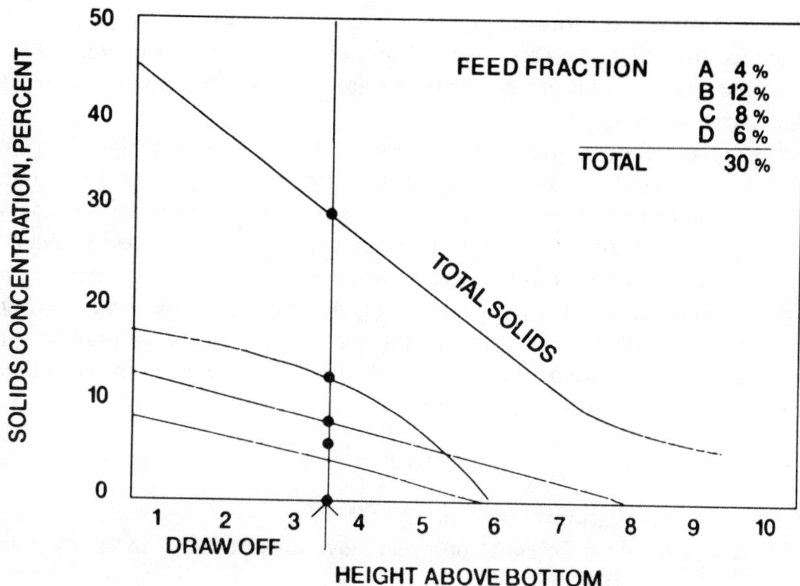

FIGURE 2-19. Solids concentration profile at various depths for 30% slurry and drawoff at one-third height. (*Mixing Equipment, Inc.*)

of the drawoff pipe at the bottom as in Fig. 2-18, or at one-third of the tank height as in Fig. 2-19, the distribution of each fraction along the height of the tank must be different (except for fraction "D", which is equally distributed in both cases, being the finest grind) to meet the requirement. With drawoff tube located at the upper position, the total weight percentage of solids under steady state conditions in the tank is increased at the expense of more horsepower. Here a balance between mixer size, operating cost and process requirements is necessary.

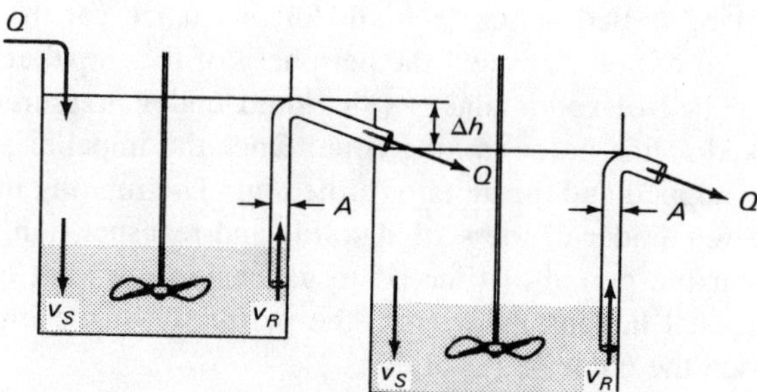

FIGURE 2-20. Continuous flow of slurry in two tanks in series utilizing risers for discharge.

Proper upcomer or drawoff pipe is illustrated as shown in Fig. 2-20. For continuous flow operations the depth of immersion of the riser is predicated on the particle size distribution, power of the agitator and elevation of suspension of the coarse grind size.

Not only is depth of immersion important for the drawoff pipe, but cross sectional area and hence rise velocity in the upcomer is important. For obvious reasons upcomer rise velocity v_R must be greater than the terminal settling velocity v_s of the largest size fraction to be discharged. For an open launder arrangement, where flow discharges from one tank to the next through an open trough on the surface of the liquid, $v_R = Q/A$, where Q is the continuous flow rate of slurry into the tank and A is the cross-sectional area of the drawoff pipe. In a closed launder system, elevations of slurry in the two tanks in series create a driving force for flow and velocity:

$$v_R = Q/A + \sqrt{2g\Delta h} \tag{20}$$

where: g = gravitational constant of 32.2 ft/sec^2
Δh = height differential between elevations of slurry in the two tanks in series, ft.

In a liquid-solids suspension application, a power failure can cause mechanical problems of major proportions. Depending upon the depth of settled solids and the degree of compaction of the bed of solids, the impeller may be completely submerged in the settled solids and startup may be impossible without jeopardizing the mechanical integrity of the drive, shaft, and/or mixer impeller. Considerable damage can occur if the motor is repeatedly jogged and the impeller will not rotate in the settled solids. The resulting inertial load imparted on the agitator can cause excessive instantaneous loads along the mixer shaft and into the drive mechanism. Discussion of how to design the agitator to compensate for this event and ensure mechanical integrity during startup in settled solids will be addressed in a later chapter.

From a process perspective, to properly start up an agitator with the impeller buried in a bed of settled solids, a liquid or gas lance can be used to help "fluidize" the bed of solids around the periphery of the impeller. The lance is inserted into the bed of solids and gas or liquid under pressure is applied to loosen the packed solids in the mixing zone. Once the impeller is free to turn, lancing may be stopped and the agitator turned on. The running mixer will help erode away the remainder of the settled solids and resuspension will normally occur in a reasonable period of time. If resuspension does not occur within a few hours additional lancing away from the periphery of the impeller will be required to loosen the compacted solids.

The power drawn by the impeller depends upon the uniformity of the slurry. If solids have settled out beneath the impeller, once it is turned on, the impeller will "see" at first only the density of the liquid, not the slurry. If the solids are tightly compacted, a new false tank bottom is in effect formed. The new bottom is closer to the impeller and this changes the power draw characteristics (i.e., axial flow impellers will draw more power at closer off-bottom locations).

As solids are resuspended, the impeller operates in more and more dense slurry. During off-bottom suspension conditions (which will naturally occur before uniform suspension is achieved during resuspension), the concentration of the slurry around the impeller will be greater than the average concentration when the solids are fully suspended. This scenario will draw more power than at complete suspension equilibrium.

Monitoring power draw during resuspension of settled solids with an indicating or recording ammeter or wattmeter will provide an accurate history of what the impeller is "seeing" regarding slurry in the mixing zone. When power draw stabilizes to a reading indicative of complete suspension, the bed of settled solids has been resuspended.

Batch operations normally require about 15–25% more power for stable suspension conditions, compared with continuous flow situations. The design of continuous flow solid suspension mixing applications requires that a provision must be made in the event there is a power failure that could stop the mixer or interrupt the feed and discharge streams. The key is to prevent solids from settling in the tank. This can be accomplished by using air lancing or recirculation by a pump, assuming the power outage does not affect compressed air or the pump or if backup power is generated to run the air compressor or pump.

In continuous flow, the entire contents of the tank need not have the same composition as the feed and exit streams. The only process requirement is to have the slurry composition at the drawoff point be constant. Therefore, the agitator should only be designed to provide and maintain uniform suspension and composition up to the discharge point. The contents above the discharge can be of different composition. In this manner, the mixer will not be oversized. Applying successively larger amounts of mixer power will make the tank contents uniform.

Slurry specific gravities may be determined by use of the nomograph shown in Fig. 2-21. When insoluble constituents are mixed, the volume of the mixture is equal to the sum of the individual volumes:

$$\frac{W}{(62.4)\ \mathrm{Sg}_M} = \frac{W_1}{(62.4)\ \mathrm{Sg}_1} + \frac{W_2}{(62.4)\ \mathrm{Sg}_2} + \cdots \tag{21}$$

where:

W = weight, lbs

Sg = specific gravity

M = subscript denoting mixture

1, 2, etc. = subscripts denoting components 1, 2, etc.

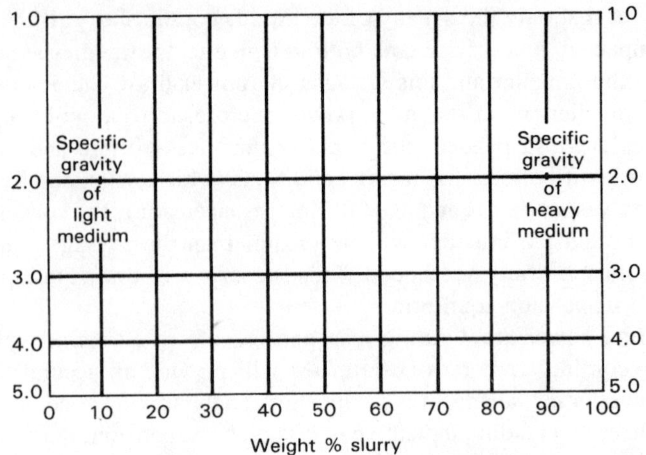

FIGURE 2-21. Average specific gravity of slurries or mixtures of immiscible liquids.

In the usual case of two components, this equation can be rearranged to give:

$$Sg_M = \frac{W_1 + W_2}{\dfrac{W_1}{Sg_1} + \dfrac{W_2}{Sg_2}} \tag{22}$$

The amount of either component is often expressed as a weight percentage P:

$$P = \left(\frac{W_1}{W_1 + W_2}\right) 100 \tag{23}$$

$$(100 - P) = \left(\frac{W_2}{W_1 + W_2}\right) 100 \tag{24}$$

Inverting both sides of Eq. (21), we get:

$$\frac{1}{Sg_M} = \frac{W_1}{(W_1 + W_2)Sg_1} + \frac{W_2}{(W_1 + W_2)Sg_2} \tag{25}$$

By rearranging Eqs. (23) and (24) and substituting these into Eq. (25), we can get:

$$\frac{1}{Sg_M} = \frac{P}{Sg_1(100)} + \left(1 - \frac{P}{100}\right)\frac{1}{Sg_2} \tag{26}$$

which rearranges to:

$$\frac{1}{Sg_M} = \frac{1}{Sg_2} + \frac{P}{100}\left(\frac{1}{Sg_1} - \frac{1}{Sg_2}\right) \tag{27}$$

Equation (27) allows for a quick solution by means of a simple nomograph on semilog paper, shown by Fig. 2-21. An operating line is drawn between the specific gravities of the light and heavy medium. A line is then drawn up from the weight percent of the slurry to the operating line and horizontally across to provide the specific gravity of the slurry.

The value of slurry specific gravity may be used in power consumption calculations and mixer design.

G. DISSOLVING OF SOLIDS

In mixing applications, dissolving usually refers to the dissolving of a solid in a liquid. The driving forces for dissolution are twofold—the adequate suspension of solids and provision of high flow rate (velocity) of liquid across the surface of the solids.

In general, for readily soluble crystalline materials, the degree of agitation that provides initial wetting and suspension of all solids will satisfy the mixing requirement for dissolution. Wetting of solids is a condition whereby solids that would normally float on the surface are incorporated or drawn into the liquid by the action of the agitator. Turbulence or vortexing (whirlpooling) on the surface of the liquid will promote wetting out of solids. For those cases where solids are difficult to dissolve or where faster dissolving rates are desired, high power levels are required.

A unique dissolving problem is often encountered when the solids are non-crystalline materials such as natural and synthetic rubbers, solid resins, and other commercial polymers. These materials initially soften and become quite sticky. The particles tend to agglomerate into larger masses and to adhere to the vessel walls, baffles, mixer shafts, etc. Very often the agglomerates will take on the appearance of "fish eyes," as the difficult to wet out solids become encapsulated with a viscous coating of the solution.

The solution usually increases in viscosity as more solids are dissolved, with final solution viscosities becoming extremely high (often greater than 500,000 centipoise) in solution having high solids content (usually exceeding 60% by weight dissolved solids). Dissolving applications of this type must take into account the viscosity factor as an inherent part of the dissolving problem. It is not uncommon to have input mixing power levels of 100 hp or more per 1,000 gallons of solution to achieve the proper balance of flow and shear for dissolving of solids. The balance weighs more heavily in favor of greater shear than flow for these mixing requirements. Where longer dissolving times can be tolerated, lower horsepower levels can be used. Actual time will depend upon the materials involved and their characteristics such as tendency to agglomerate, stickiness of the solids, ability of solids to wet out, etc., in addition to temperature, particle size, and liquid viscosity.

Examples of various dissolving applications are presented below.

Easy Dissolving Application

An easy dissolving application is salt solution makeup, such as preparing a 20% brine (sodium or calcium chloride) solution in a 2,000 gallon tank at temperatures of 150°F. This will require a one horsepower mixer with a constant pitch axial flow turbine, based on off-bottom suspension of salt for dissolving.

Medium Difficulty Dissolving Application

Dissolving 1% carboxymethyl cellulose in 2,000 gallons of water, where final viscosities could run as high as 7,500 centipoise, will require 7½ to 10 hp with a constant pitch axial flow impeller. Starch cookers using turbine mixers are a good example of a mixing application that requires both solids dispersion and dissolution. At a point in the cooking cycle the starch changes from a slurry to a viscous solution and passes a peak viscosity before it is finally diluted. The mixer must be designed for the peak viscosity and the controlling factor in the selection of the agitator is viscous blending rather than solid suspension.

For example, a 1,000 gallon starch batch at 12,000 centipoise peak viscosity would require a 3 hp mixer. At 80,000 centipoise peak viscosity during cooking, 7½ hp would be required.

Extremely Difficult Dissolving Application

An example of a difficult dissolving application is rubber crumb dissolving in organic solvents. Final viscosities can easily reach 500,000 to 1,000,000 centipoise or higher depending upon solids content. Dissolving rates can be accomplished slowly over several hours at moderate horsepower inputs with large diameter slow speed impellers producing a greater balance of flow than shear. Power levels in the range of 25 hp per 1,000 gallons are typical.

Dissolving time can be shortened by up to 80% by sharply increasing power input and providing higher shear in addition to high fluid turnover. Power levels in this instance will be in the range of 200 hp per 1,000 gallons.

H. DRAFT TUBE CIRCULATORS

Another efficient use of power for solids suspension is draft tube circulation. This is the most economical way to use power in solid suspension applications.

A draft tube is a large diameter, open-ended pipe, suspended in a tank in such a way that the entire tube is submerged. The mixing impeller runs inside the draft tube, pumping up or down, creating velocities along the tank bottom and in the annular space to maintain suspension. Generally, the mixer will pump down the draft tube and up the annular space. The annular rise velocity required is usually 5–10 times the terminal settling velocity of the *largest* and or *heaviest* particle. Since the rise velocity must be maintained over a smaller annular area with a draft tube circulator, compared to an open impeller tank, the overall pumping rate and therefore power requirements are significantly less with the draft tube. A rough rule of thumb is that draft tube circulators require 30–50% of the power requirements of an open impeller tank to achieve similar levels of solids suspension. Actual power savings is dependent on whether the slurry exhibits free or hindered settling characteristics, diameter of draft tube, type and diameter of impeller, impeller location, draft tube submergence and off-bottom location, etc.

Up-pumping draft tube circulators draw solids in from the annulus into the suction side of the tube. The ideal tank geometry for this mode of mixing calls for conical or deep dished bottoms. Further, there must not be too high a solids concentration in the annulus or the particles will "crowd" themselves into higher concentrations while trying to enter the draft tube.

Disadvantages to draft tube circulators include the added capital cost associated with the tube and special design provision required to resuspend after a power failure. To encourage resuspension of settled solids following a power outage, the draft tube must be designed with vertical slots rising up from the bottom of the tube to a height above the predicted level of settled solids. The slot area must be designed to provide adequate velocity of supernatant liquor through the slots and across the bed of settled solids to promote erosion of the solids and ultimate resuspension. The restriction of flow out the bottom of the tube and redirection through the slots must be generated against significantly increased resistance or head in the system. This requires greater power than during normal operation. So it is imperative that the draft tube circulator mixer be designed for greater power and head requirements to be experienced during resuspension of solids following a power failure. To do this, one must generate the head–flow curve for the draft tube circulator mixer in question and determine the operating point on this curve during resuspension.

The draft tube mixer per se may be more expensive to buy than an open impeller agitator due to the special hydrofoil impellers used in draft tube circulators.

Draft tube circulators can only operate in constant or nearly constant liquid levels in the tank. Obviously the liquid level must always remain above the top of the draft tube for circulation to occur. Thus a constant slurry flow or continuous operation is needed, with inlet and outlet design and location being very critical to avoid "short circuiting" of incoming material across the liquid surface and out the discharge and to maintain proper liquid levels in the tank.

"Interparticle communication" is not as good in a draft tube circulator as it is in an open-impeller tank. Also draft tubes are more complicated to install and maintain and are usually limited to moderate-power systems.

In continuous flow slurry suspension application with a draft tube circulation, further mixer power and capital cost savings can be achieved by suspending the solids only part way up the annulus. The outlet pipe would have to run down to the level of the suspension. The outlet riser has to be extended down below the level of suspension and the slurry composition must be uniform at and below this point. In fact, the composition must be the same as the inlet. Care must be taken to minimize resistences to flow (i.e., increases to head) in extending the outlet riser down. The riser diameter should be calculated based on continuous flow rate into and out of the tank, and a riser velocity at least 5–10 times the terminal settling velocity of the *largest* or *heaviest* particle.

Draft tube circulators were developed in the alumina industry for trihydrate crystallizers. Draft tube systems have been used in other mineral processing solid suspension applications such as uranium, copper, gold slurry leach tanks, etc. Optimum draft tube diameter is critical to minimize power requirements. The overall circulator system must be designed as an axial flow pump operating in a closed circuit. A draft tube mixing system exhibits characteristic behavior requiring a variable pitch, hydrofoil impeller for best hydraulic stability and efficiency. Tank height may be two or more times diameter.

Resuspension of settled solids can be assisted by a series of slots in the draft tube. These slots have the secondary effect of stabilizing hydraulic behavior and reducing installed power requirements.

Overall, draft tube circulators will be higher in capital cost than open impeller mixers, but can be 50–70% lower in power costs.

Traditional air-lift "pachuca" mixing systems preceeded the mechanically agitated draft tube circulator in minerals processing slurry mixing applications, such as leach tanks. With rising energy costs, this type of mixing became inefficient, often requiring up to ten times more power to mix with compressed air in a tube than with a hydrofoil mechanically mixed draft tube circulator.

Attempts have been made to provide mechanical circulation by installing a down pumping impeller in a "pachuca" tube. Although such an arrangement may work after a fashion, and even give some power savings, it is not optimum because the draft tube diameter is too small. Further, it is very difficult to suspend solids from the apex of the "pachuca" tank cone, so either the tank bottom must be modified or supplementary agitation or pump out must be provided to deal with these solids.

Superficially a draft tube circulator resembles a retrofitted "pachuca." However, draft tube, tank, and impeller geometry have been optimized to give the required flow with minimum shear. Shear stresses due to agitation are significantly less than with open turbine mixers. For this reason, draft tube circulators work well with shear sensitive slurries and crystals. Circulator size and mixer power are essentially independent of tank height, so in tall tanks, power requirements per unit volume can be reduced by 50–70% if an efficient hydrofoil and draft tube are incorporated. The extra cost of a circulator installation is generally justified by power savings and superior performance.

Draft tube diameter is predicated on two independent velocity requirements, both of which must be satisfied to achieve uniform suspension of solids.

Velocity down the draft tube, V_D, is set by the requirement to sweep the bottom of the tank and prevent solids deposition. The draft tube velocity depends on percent solids, solids specific gravity, particle size distribution, and slurry viscosity—shear characteristics. V_D must be determined experimentally and as accurately as possible.

The rise velocity in the annulus, V_a, must be sufficient to suspend uniformly, and circulate the slurry. It is a multiple of the settling velocity of the largest particle in a free settling scenario—usually 5–10 times. Although many slurries exhibit hindered settling characteristics, the design criterion is predicated on free settling conditions typical of that experienced at low solids concentrations during process upsets or start-up.

The optimum draft tube diameter D_T is sized to accommodate and satisfy both velocity requirements.

In Fig. 2-22 the power required to produce these velocities is plotted against draft tube to tank diameter ratio. Power requirements are at a minimum where these two curves intersect. This occurs at:

$$D_T/T = \sqrt{\frac{V_a}{V_a + V_D}} \tag{28}$$

where: T = tank diameter
D_T = draft tube diameter.

Approach to Design—Draft Tube Circulator

A circulation impeller in a draft tube behaves like an axial flow pump of high specific speed. The draft tube circulator must be designed on the same basis as any other pump system. The operating point is where the head versus flow curve for the pump intersects the resistance head versus flow curve for the circulated fluid, as shown in Fig. 2-23. But behavior must also be examined for startup and process upset conditions.

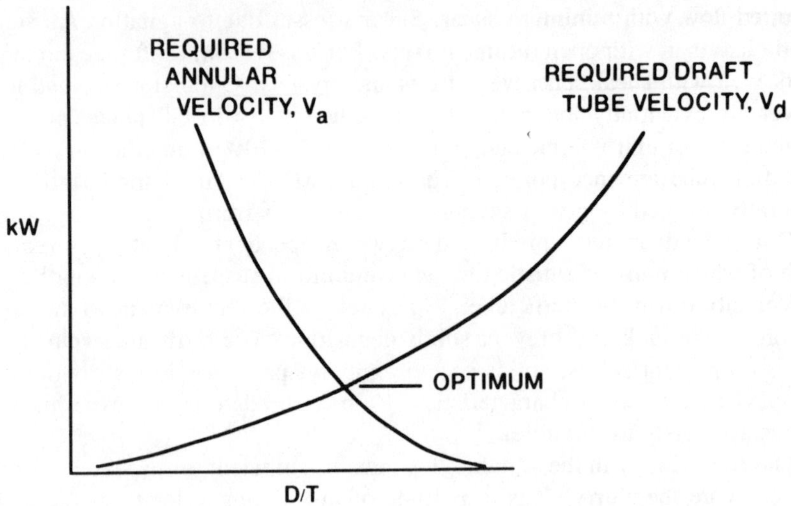

FIGURE 2-22. Energy optimization in the design of draft tube circulators. (*Mixing Equipment, Inc.*)

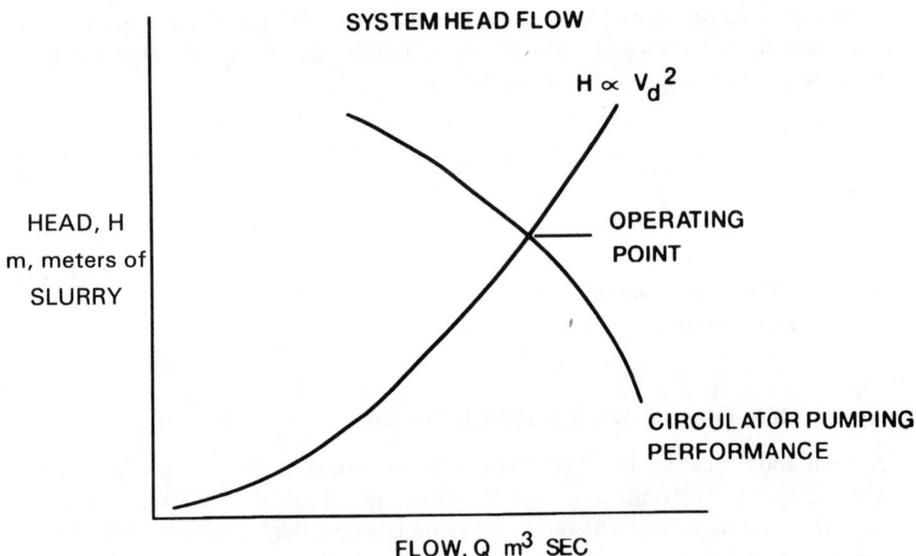

FIGURE 2-23. Pumping performance in a draft tube circulator. (*Mixing Equipment, Inc.*)

System Head Versus Flow

The system head or resistance to flow is made up of several components of head loss due to:

1. Draft tube entrance contraction of flow
2. Draft tube internals
3. Draft tube exit expansion of flow
4. Impingement of flow on tank bottom and change in direction of flow
5. Frictional loss in draft tube and annulus

Frictional losses in the annulus are infinitesimal and friction losses in the draft tube are negligible at slurry viscosities below about 1,000 centipoise. Careful design can minimize loss due to draft tube entrance and internals. Beveled inlet and outlet transition pieces can streamline flow patterns in the top and out the bottom of draft tubes, thus minimizing pressure drop in the system. The circulator impeller can operate in a notched section of the tube, in such a way that the impeller "thinks" it is running in close clearance to the tube wall. See Fig. 2-24 for an illustration of optimum draft tube configuration. An axial

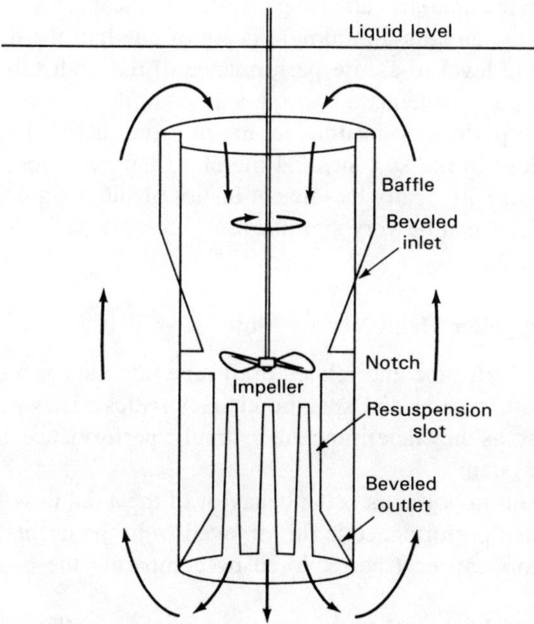

FIGURE 2-24. Optimum draft tube configuration.

flow pump running in close clearance to its housing inherently maximizes flow efficiency. Clearance should be 1–2″ from the notch wall to blade tip. Flow reversal losses are the most significant component although the placement of outlet launders or riser tubes must be done with care to avoid introducing additional head loss at the top of the tank.

Determination of system head loss is very important and cannot be predicted accurately enough from hydraulic theory. It must be measured in fully instrumented test tanks of significant size to avoid serious Reynolds number discrepancies on scaleup.

It is often convenient to express system head loss as a *K*-factor (i.e., a multiple of velocity heads in the draft tube). A draft tube of sophisticated design, with beveled inlet and internal notch, operating at high Reynolds number can have a *K* as low as 1.25, although 1.4–1.5 is typically assumed for higher solids slurries. A primitive draft tube design without beveled inlet and internal notch may have *K* factor of 2.0 or more.

During a process upset, the system head versus flow curve may be modified considerably. For example, a partially obstructed draft tube exit will give a much steeper curve with a sharp increase in *K* factor. Low liquid level can cause head loss across the draft tube entrance and superimpose a static head differential. Quite small differences in specific gravity between the draft tube contents and material in the annulus can also generate significant static head differentials. It is desirable to maintain a minimum depth of one-half the draft tube diameter below the liquid level to assure performance of the draft tube circulator. Low liquid levels create inadequate suction head conditions, much in the same way a pump fails to perform at insufficient net positive suction heads.

The head loss in the system, and therefore the head developed by the circulator impeller, is typically less than 4 inches of slurry pressure drop, so small static head effects can be very significant.

Circulator Impeller Head Versus Flow

The impeller, draft tube and related draft tube internals as a system is an axial flow pump with special and specific characteristics. These characteristics are very important, as they determine the hydraulic performance and power requirements of the system.

Of significant importance is the behavior of the axial flow pump under upset conditions when performance is far removed from the normal operating point.

These factors can best be explored by comparing the behavior of two impellers:

Type A is a 4-bladed, flat plate, constant pitch axial flow turbine with blades angled at 30–35°, running in a simple draft tube with a beveled conical inlet.

Type B is a hydrofoil turbine with cambered blades and a progressively increasing pitch angle from tip to hub. The draft tube has a beveled entrance, inlet baffles, stator vanes below the impeller to help straighten the flow away from the impeller. Tip recirculation is suppressed by a circumferential notch opposite the impeller.

Typical head versus flow performance curves for these two impellers are shown in Fig. 2-25. There are significant differences in performance and efficiency.

Impeller Behavior

With many respects the hydrofoil impeller blade can be regarded as analogous to an airplane wing. An important consideration is the "angle of attack." This is the angle between the centerline of the blade section and what the blade perceives to be the direction of flow of liquid across it. To put it another way, the angle of attack is the angle that the resultant axial and radial velocity vector makes with the trailing edge of the blade.

As flow decreases, the vertical (axial) velocity vector decreases while the horizontal (radial) velocity component remains constant. Fig. 2-26 shows how as flow decreases, the flow vector relative to the impeller becomes more horizontal and the angle of attack of the blade increases. Different response to changing angle of attack explains the differences in impeller behavior.

In the operating region, the Type B impeller has a much steeper curve than

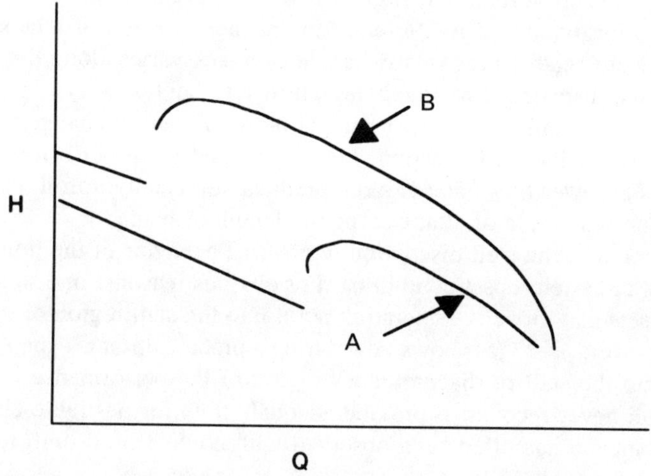

FIGURE 2-25. Impeller head versus flow characteristics. (*Mixing Equipment, Inc.*)

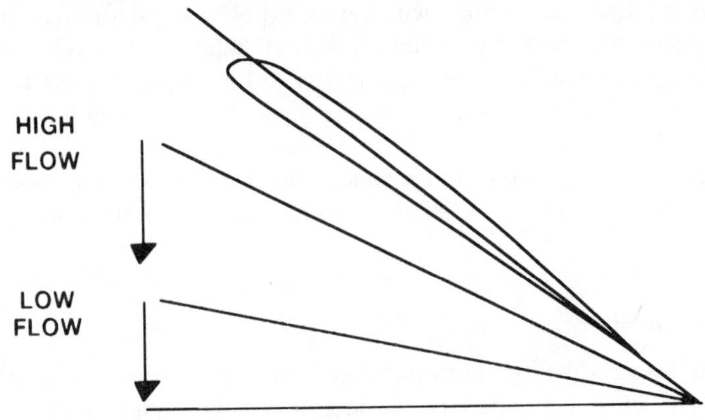

FIGURE 2-26. Angle of attack. (*Mixing Equipment, Inc.*)

Type A because at higher angles of attack a hydrofoil generates more lift (i.e., head) than a flat plate, constant pitch turbine. The hydrofoil is more forgiving to increases in head in the system. Flow reduction is minimized due to the steepness of the performance curve. As one moves back on the performance curve to another operating point, there is proportionately a much lower reduction in flow than increase in head.

At lower flows there is discontinuity in both curves. This is a characteristic of all axial flow pumps. It corresponds to the stall point of an airfoil where boundary layer separation and turbulence occur, lift is lost and head falls off sharply. This occurs at relatively high flow with a constant pitch turbine because (a) it is less forgiving and less tolerant to changes in angle of attack with increasing system head (b) the relative angle of attack varies along the length of the blade. Stall happens at relatively high flow rates and is fairly unpredictable. The Type B hydrofoil on the other hand is more tolerant to change in angle of attack because of its airfoil section, blade twist and steep performance curve. Stall occurs at lower flows and is more predictable. The hydrofoil has variable pitch but constant angle of attack along the length of blade.

The location of the stall discontinuity governs behavior of the impeller during process or system upset conditions. For obvious reasons, one should avoid process upsets that move the operating point into the stall region of the performance curve. Fig. 2-27 (a) shows how during a process upset a Type A impeller may fall into the stall or discontinuous region of the performance curve from which it can never recover to provide adequate flow for desirable circulation. This phenomenon has often been observed with nonhydrofoil draft tube circulator impellers.

In contrast, Fig. 2-27 (b) shows how a Type B system with hydrofoil impeller is stable over a wide operating range, is more forgiving to process upsets,

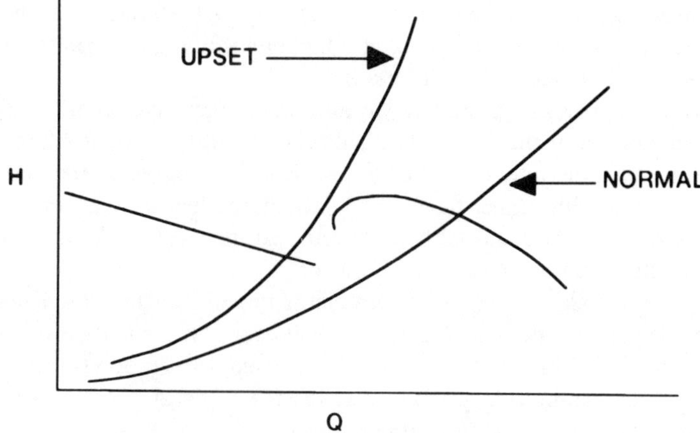

(a) TYPE "A" FIXED PITCH VARIABLE ANGLE OF ATTACK AXIAL FLOW TURBINE

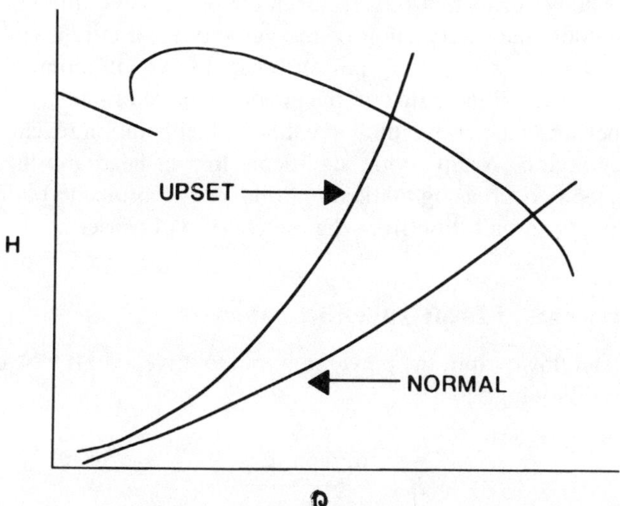

(b) TYPE "B" VARIABLE PITCH CONSTANT ANGLE OF ATTACK HYDROFOIL

FIGURE 2-27. Upset conditions in draft tube circulators. (*Mixing Equipment, Inc.*)

and is capable of overcoming and recovering from even major process upsets. It does this without shifting the operating point into the stall or discontinuous region of the performance curve.

Process upsets occur when the following happens:

1. Solids concentration decreases, thereby reducing the hindered settling affect of the solids (i.e., increasing the settling velocity of the larger particles).

2. Large size particles tend to accumulate in the bottom part of the annulus of the circulator tank, creating a slightly higher specific gravity in this area with a resultant smaller static head differential.

3. Solids deposit along the tank bottom in a progressive manner building up toward the bottom of the draft tube. This chokes off the flow discharging from the tube, moving the intersection of the velocity head curve up the performance curve to a point with higher head development and less flow generation.

4. Viscosity of the material being circulated increases creating a greater resistance to flow due to increased viscous drag.

5. Gas is introduced below the circulator impeller or increased in flow beneath the impeller. The gas creates a buoyant force that must be overcome by the circulator impeller. Draft tube circulators used for aeration of waste waters, slurry leach or gas-liquid, gas-liquid-solids reactions in continuous flow scenarios are prone to this type of process upset.

If any of the above process upsets result in the progressive filleting of solids along the tank floor, the intersection of the velocity head curve with the performance curve (i.e., the operating point) moves farther up along the performance curve. As more solids settle out, the operating point or intersection moves farther up the performance curve, until eventually the operation reaches the stall or discontinuity region. At this point significant loss of head and flow is seen with annular velocity decreasing markedly. Solids will continue to build up until the draft tube plugs and all flow from the circulator will cease.

Power Requirements of Draft Tube Circulators

Similar to an axial flow pump, the power required to drive a draft tube circulator is given by the following:

$$P = \frac{9.788\ QH\rho}{\eta} \tag{29}$$

where: P = power required, kilowatt (kW)
Q = flow produced, cubic meter per second (m^3/sec)
H = head developed, meters of fluid flowing (m)
ρ = specific gravity of fluid flowing
η = hydraulic efficiency.

From this equation it can be derived that:

$$P \propto V_D^3 \tag{30}$$

So accurate experimental determination or measurement of V_D, draft tube velocity, is very important.

Hydraulic efficiency η for a given hydrofoil geometry may be determined by running test conditions with an experimental circulator. Head and flow can be measured with devices such as ottmeters and kiel probes or pitot tubes, etc. More accurate data may be had with laser velocimeters and computer integration of velocity over the sectional area of the draft tube. Ammeter or wattmeter recordings can provide measurements of power draw.

By running a series of tests with different impeller geometries, at various speeds, with and without controlled resistances to flow, and measuring variables such as flow (velocity), head and power, one can determine hydraulic efficiency and generate performance curves.

Let us consider the following example of a design calculation for a draft tube circulator.

A tank 12 feet in diameter and 24 feet high is to be mixed with a draft tube circulator.

The draft tube diameter will be 4 feet and a hydrofoil with a power number, $N_p = 1.0$, will be used. The mixing application requires a circulation rate of 20,000 gpm and performance of this pumping rate must be maintained against a head of 1.0 meter (i.e. based on pumping 20,000 gpm through a 4 foot diameter tube, the head requirement is 1.0 meter).

The specific gravity is 1.0 and hydraulic efficiency of the hydrofoil is 0.67. Converting gpm to cubic meters per second:

$$Q = \left(\frac{20{,}000 \text{ gallons/min}}{252 \text{ gallons/m}^3}\right)\left(\frac{1 \text{ min}}{60 \text{ sec}}\right) = 1.323 \text{ m}^3/\text{sec}.$$

From Eq. (29),

$$P = 9.788(1.323)(1.0)(1.0)/0.67 = 19.33 \text{ Hp}.$$

Circulator mixer speed may be calculated as follows: Assume for a 4 foot draft tube diameter, the impeller diameter will be 44 inches. From Eq. (4):

$$P = N_p \rho N^3 D^5 / 1.53 \times 10^{13}$$

$$19.33 = (1.0)(1.0)(N^3)(44^5)/1.53 \times 10^{13}$$

Rearranging and solving for N,

$$N = 122 \text{ rpm}.$$

Resuspension of Solids in Draft Tube Circulators

Tapered slots or resuspension vents in the lower part of the draft tube can help resuspend settled solids. See illustration in Fig. 2-24. If, after a power failure, the end of the draft tube is blocked off by settled solids, flow passing through the exposed section of slot at high velocity will progressively erode away at the layer of solids and ultimately resuspend the bed of solids. As the bed of solids is continuously swept by flow through the resuspension vents, the operating point moves down the performance curve until the bed is completely resuspended and a normal operating point is achieved.

The resuspension slots have an important effect on the behavior of the system. Because there is always some bypass of flow through the vents, even during normal operation, a complete shutoff, zero flow condition is never experienced. Effectively, this does away with the unstable low flow, high head area on the performance curve. Indeed, with Type B hydrofoil impellers the discontinuity may be completely eliminated, giving a stable characteristic over the whole operating range. See Fig. 2-28(a) for the effect that slot bypass has on stabilizing the performance of the draft tube circulator under low flow conditions.

In a like manner, the power versus flow curve, which rises sharpely at low flows (due to high head development and/or poor hydraulic efficiency), is stabilized by flow bypass through the vents. A drive motor and speed reduction drive assembly sized for twice normal operating power would be required with-

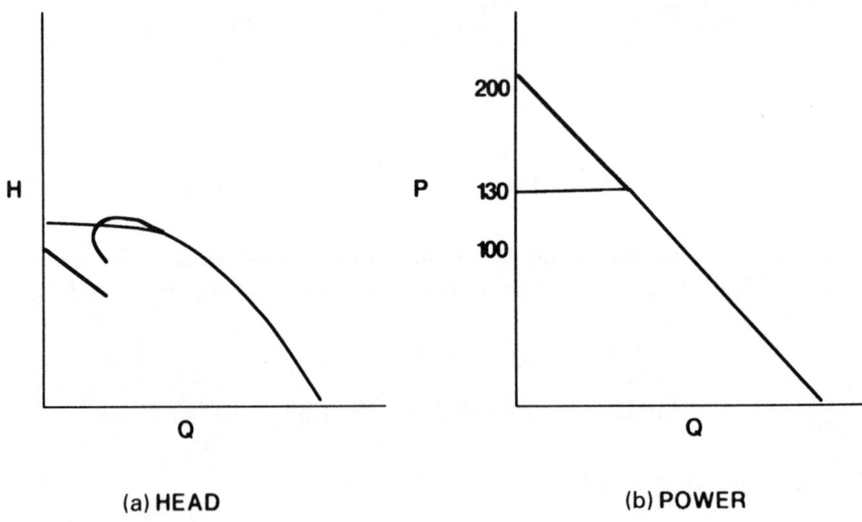

FIGURE 2-28. Slot bypass effect. (*Mixing Equipment, Inc.*)

out bypass through the resuspension vents. With flow bypass through the slots, an over capacity of only 33% is required for power. This is demonstrated by Fig. 2-28(b).

Slurries that are hindered in settling velocity and settle quite slowly, behave in a similar fashion to free settling slurries in a slotted draft tube system, if they contain a proportion of larger particles that still concentrate fairly quickly. This is often the case in minerals processing slurries that are predominated by fine grind fractions and of significant solids concentration (i.e., 80% of all solids finer than 200 mesh and 45–55% by weight solids).

The choice between open impeller mixing and draft tube circulation is primarily economic, a tradeoff between capital and operating cost factors (i.e., installed cost of tank and mixer versus mixing power consumption and maintenance cost).

Tank geometry is also a factor and may be set by site conditions, other process requirements, or by the need to retrofit the mixing system into existing tankage. Generally speaking, open impeller agitators are better suited for squat tanks where height does not exceed the diameter. A draft tube circulator is most effective in a tank height/diameter ratio in the range of 1.5–3.0.

Given that tank geometry is optimized with respect to height/diameter, a properly designed circulator mixer operating in a draft tube can reduce power by 50–70% over that of open impeller mixers. However, draft tube circulators are expensive to install. The hydrofoil impeller can be very sophisticated in design and expensive to fabricate. The draft tube and support structure may cost as much as the circulator mixer. Large tank bottom fillets are required in circulators although wall baffles are not.

Draft tube circulators must operate with a full tank and nearly constant liquid level whereas open impeller agitators can run with a wide variation in liquid level.

The open impeller mixer is prone to "sanding in" in the event of a power outage and must be designed mechanically to deal with this situation without

TABLE 2-5. Comparative Analysis—Draft Tube Circulator and Open Impeller Mixer.

Best Tank Shape	Draft Tube Circulator Z/T = 1.5-3.0	Open Impeller Mixer Z/T = 0.8-1.0
Draft tube	Yes	No
Wall baffles	No	Yes
Filleted bottom	Yes	No
Capital cost	High	Lower
Power cost	Lower	High
Torque	Lower	High

T = diameter, Z = height.

damage to the unit. The draft tube circulator does not have this problem, and with a slotted draft tube easily resuspends solids after a power failure.

Some of the more important and pronounced differences are shown in Table 2-5.

I. HEAT TRANSFER

Heat transfer applications require flow controlled mixing design criteria, as opposed to shear controlling considerations. Applications include agitation to provide heating, or cooling, or merely to maintain a uniform temperature of the tank contants. Maintaining temperature uniformity is the easiest mixing application. Heating is the next easiest application, followed by cooling as the most difficult.

In processing, heat is transferred by conduction from the wall of the vessel or surface of the internal heating elements (i.e., coils) to the contents of the vessel. As the contents are agitated, heat is transferred into the mass by convection. Mixing thereby speeds up the conduction of heat from the heat transfer surfaces, and helps promote the transfer of heat by forced convection.

Heating and cooling applications are best handled by providing adequate flow velocities of fluid across the heat transfer surfaces, with appropriate measures taken to promote good top-to-bottom circulation and turnover of tank contents. This flow pattern will assure good exchange of fluid from areas remote to the heat transfer surface to the heat transfer element (i.e., between fluid in the center of the tank and the jacketed tank wall). In this manner reasonably uniform temperature profiles will result.

Heat transfer mixing applications may be considered similar to blending requirements. Consequently, for low viscosity liquids, conventional propeller, impeller, and state-of-the-art hydrofoils are generally used. One qualification on horsepower levels for low viscosity applications is that unlike blending, a point of diminishing returns can be reached for heat transfer where an increase in input horsepower has an incrementally diminishing effect on process result.

For higher viscosity liquids, larger diameter, slower speed impellers are usually required to obtain the required flow patterns. Often multiple large diameter, slow speed turbines are needed with accompanying increases in horsepower similar to that required in high viscosity blending applications.

Complete data on all contributing factors in a heat tranfer agitator application are often difficult to obtain. As much data that is available should be furnished, including the following:

- Tank dimensions and details of heat transfer surface (i.e., jacket-type—half pipe, dimple, etc.; surface area; coil pipe or tube size and wall thickness; number of banks of coils and diameter of each bank; coil spacing; etc.)

- Whether heating, cooling, or temperature uniformity is required process result
- Time available for heating and/or cooling
- Are solids present so that solid suspension must be addressed? If so, details must be provided as in any solid suspension problem to design a mixer to meet this criteria
- Viscosity and specific gravity of fluids at the temperatures to be encountered
- Details on heating or cooling media and temperature of batch at the start and end of the cycle
- Specific heats and thermal conductivities of tank contents
- Fouling characteristics of the heating or cooling medium and is the process material heat sensitive (i.e., will it decompose at the temperature of the heating medium? Can it solidify at the temperature of the cooling medium?)

Agitation to improve heat transfer can be classified into two broad categories: (a) open impellers running at some distance from the tank wall and (b) impellers operating at close clearance to the tank wall, typically at clearance less than 5% of the tank inside diameter. Examples of open impellers are propellers, radial and axial flow turbines of constant pitch, and hydrofoils of variable pitch. Close clearance impellers are helical and anchor turbines. In all cases, the impellers provide a greater balance of flow than shear to enhance heat transfer.

Close clearance impellers are often used with highly viscous liquids, for which homogeneous mixing is critical in controlling say heat of reaction, heat of crystallization, etc. Open impellers are not suitable in high viscosity heat transfer applications because they cannot develop or maintain complete motion throughout the tank volume, especially at places remote from the impeller (i.e., at tank wall).

The transfer of heat to and from an agitated tank depends on the temperature-difference driving force (ΔT) between the heat transfer medium and process fluid, the heat transfer area (A_o) and the overall heat transfer coefficient (U_o):

$$Q = U_o A_o \Delta T \tag{31}$$

where: Q = heat transfer rate, BTU/hr
U_o = overall heat transfer coefficient BTU/hr-ft^2°F
A_o = heat transfer area, ft^2
ΔT = temperature difference between heat transfer media and tank contents, °F

The calculation of ΔT and A_o is straightforward, but U_o is a combination of several resistances to heat transfer. Thus,

$$\frac{1}{U_o} = \frac{1}{h_o} + \frac{l}{k} + \frac{1}{h_i}\frac{A_o}{A_i} + \text{ff} \tag{32}$$

The subscript o refers to the process side or mixer side and the subscript i to the heat transfer medium side. The overall heat-transfer resistance $(1/u_o)$ is the sum of the individual resistances due to the outside film coefficient on the mixer side of the heat transfer surface $(1/h_o)$, the wall material and thickness (l/k), the inside film coefficient of the heat transfer medium $(1/h_i)$ corrected to the common heat-transfer area basis, and a fouling factor (ff) to account for corrosion, dirt films and scale formation on either side of the heat transfer surface.

The agitator can only affect the mixer side heat transfer film coefficient (h_o). In many cases, the mixer side film coefficient will be the governing resistance, and therefore, the effect of a mixer on the overall heat transfer coefficient will usually be quite significant.

To predict heat transfer rates in agitated tanks the mixer side heat transfer film coefficient (h_o) must be estimated. By means of dimensional analysis of heat-flow and energy-balance equations, the Nusselt number (which contains the coefficient h_o) can be expressed as a function of Reynolds number and Prandtl number:

$$N_{Nu} = f(N_{Re}, N_{Pr}) \tag{33}$$

where: N_{Nu} = Nusselt number, dimensionless
N_{Re} = Reynolds number, dimensionless
N_{Pr} = Prandtl number, dimensionless.

Equation (33) can be expanded into a form similar to the heat-transfer relationship dependent on impeller characteristics, geometrical properties of the tank, and fluid properties:

$$N_{Nu} = \frac{h_o D}{K} = f\left(\frac{D^2 N \rho}{\mu}, \frac{C_p \mu}{K}, \mu/\mu_s, \frac{D}{T}, \frac{d}{T}\right) \tag{34}$$

where: D = impeller diameter
K = *thermal conductivity of tank contents*
N = impeller speed
ρ = density of tank contents
μ = viscosity of tank contents

μ_s = viscosity of tank contents at the temperature of heating or cooling medium (i.e.: at the heat transfer surface)
C_p = specific heat of tank contents
d = diameter of coil pipe or tubing
T = tank inside diameter.

The relationship expressed in Eq. (34) includes many independent variables in the form of dimensionless groups (i.e., $N_{Re} = D^2N\rho/\mu$ and $N_{Pr} = (C_p\mu/K)$ but only a portion of those that affect the flow field and heat transfer coefficient. All the variables can be broken down into three categories: geometrical properties, impeller characteristics and fluid properties. For a vertical cylindrical tank, the following are the principal variables for these three classes:

1. Geometrical properties:
   Tank diameter, T
   Liquid depth, Z
   Bottom geometry (flat or dished, etc.)
   With or without baffles (number of baffles $= \eta$)
   Ratio of impeller to tank diameter (D/T)
   jacketed, helical coil, vertical panelcoil, or vertical tube heat transfer surface
   Coil or tubing spacing, S
   Number of coils or tubes
2. Impeller characteristics:
   Type (axial or radial flow turbine, hydrofoil, etc.)
   Diameter, D
   Number of blades
   Blade geometry (height, width)
   Impeller off-bottom and spacing (if multiple turbines)
   Impeller blade pitch
   Rotational speed, N
   Placement of impeller with respect to coils
3. Fluid properties:
   Newtonian or non-Newtonian characteristics
   Viscosity as a function of temperature and shear rate
   Thermal conductivity, K
   Thermal heat capacity, C_p
   Density, ρ

The effects of many of these variables on heat transfer have never been investigated in detail as it is impossible to isolate them and quantify their effect. However, several correlations are presented below. In some cases, though, the results of experimental investigation of one heat transfer system have to be related to those of another. Because the variables in the Prandtl and Reynolds numbers have been more thoroughly investigated, their effects on mixer side heat transfer film coefficient have been fairly well established.

The three most common types of heat transfer surfaces are jacketing, helical coils and vertical tubes. These are illustrated in Fig. 2-29.

For each type experimental correlations are presented for estimating the process fluid film coefficient. The calculated coefficient can only be estimated, because many of the variables affecting heat transfer in agitated tanks are not included in the correlations. All the correlations are based on fully or substantially baffled tanks, since the industrial use of unbaffled tanks in other than high viscosity mixing applications is quite limited.

Jacketed Tanks

The jacket surface, which has the fewest variables to analyze, offers advantages with respect to initial cost and operating economy. Its principal limitations are the heat transfer surface area is small for a given tank volume, in a commercial scale application. Another drawback to jackets is that adjustment of the jacket area after the tank has been constructed is virtually impossible.

Equation (35) shows the relationship of process-side (mixer-side) heat transfer film coefficient, h_o, to geometrical and fluid properties, and impeller characteristics, for jacketed heat transfer, with a flat-blade turbine, covering a range

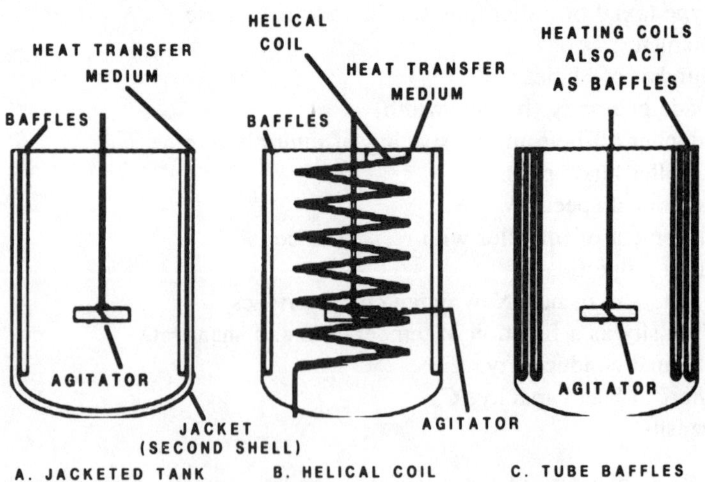

FIGURE 2-29. Three common heat transfer surfaces with agitated tanks. (*Mixing Equipment, Inc.*)

of low to moderate viscosities:

$$\frac{h_o T}{K} = 0.74 \left(\frac{D^2 N \rho}{\mu}\right)^{2/3} \left(\frac{C_p \mu}{K}\right)^{1/3} \left(\frac{\mu}{\mu_s}\right)^{1/7} \tag{35}$$

The exponents on Reynolds number, $D^2 N\rho/\mu$, and Prandtl number, $C_p\mu/K$, have been well documented for paddle, axial flow turbines and hydrofoils, and flat bladed radial flow impellers. The exponent on viscosity ratio, μ/μ_s, may range from 1/7 to 1/4, but unless the temperature differential is extremely great between the process side film on the tank wall and the tank contents, the ratio is likely to be close to unity. In such a case, the value of the exponent is inconsequential. When calculating the process-side heat transfer film coefficient, one should assume the viscosity ratio to be unity.

More recent experimental work seems to support the following correlation:

$$\frac{h_o T}{K} = 0.85 \left(\frac{D^2 N \rho}{\mu}\right)^{2/3} \left(\frac{C_p \mu}{K}\right)^{1/3} \left(\frac{\mu}{\mu_s}\right)^{1/3} \left(\frac{Z}{T}\right)^{-5/11} \left(\frac{D}{T}\right)^{1/8} \tag{36}$$

Helical Coils

A significantly greater amount of heat transfer surface can be obtained with helical coils for a given volume of process fluid than with jacketed tanks. However, coils are usually more costly to fabricate and more difficult to maintain.

The following general correlation for helical coils in baffled tanks is based on experimentation and thorough investigation.

$$\frac{h_o d}{K} = 0.17 \left(\frac{D^2 N \rho}{\mu}\right)^{2/3} \left(\frac{C_p \mu}{K}\right)^{3/8} \left(\frac{D}{T}\right)^{1/10} \left(\frac{d}{T}\right)^{1/2} \left(\frac{\mu}{\mu_s}\right)^{M} \tag{37}$$

As you can see, the exponents on the dimensionless groups are nearly identical to those of jacketed tanks, and are consistent with earlier helical-coil heat transfer investigations. An additional variable for helical coils compared to jacketed tanks is tube diameter d. Eq. (37) is based on work covering the practical and commercially available range of tube diameters. Tube spacings of between 2 and 4 tube diameters (centerline to centerline) are required for consistent results. Spacing tighter than 2 tube diameters will not allow for good flow and velocity of fluid across the tube surfaces. Spacing greater than 4 tube diameters apart minimizes coil surface area, and may require multiple banks of coils or outside surface area to compensate.

A correlation obtained for exponent M, for the viscosity correction factor (μ/μ_s) results in values ranging from 0.1 to 1.0 for M:

$$M = 0.1\ (\mu 8{,}621 \times 10^{-5})^{-1/5} \tag{38}$$

A plot of the exponent M versus viscosity for a typical fluid is shown in Fig. 2-30.

Vertical Tubes

For this style of heat transfer surface, the tubes are areas of high turbulence, improving heat transfer. The best work on vertical tubes as a surface for heat transfer has resulted in the following relationship:

$$h_o \frac{d}{K} = 0.09 \left[\left(\frac{D^2 N \rho}{\mu}\right)^{2/3} \left(\frac{C_p \mu}{K}\right)^{1/3} \left(\frac{D}{T}\right)^{1/3} \left(\frac{2}{\eta_B}\right)^{1/5} \left(\frac{\mu}{\mu_s}\right)^{1/7} \right] \tag{39}$$

where η_B is the number of vertical tubes acting as baffles.

Previous work indicates that with six tube baffles (i.e., $\eta = 6$), each consisting of three tubes of size $d/T = 0.031$, the vertical tubes account for 75% of the power consumption of an impeller in a tank having standard wall baffles. Thus, tube baffles provide reasonable, but not full, baffling.

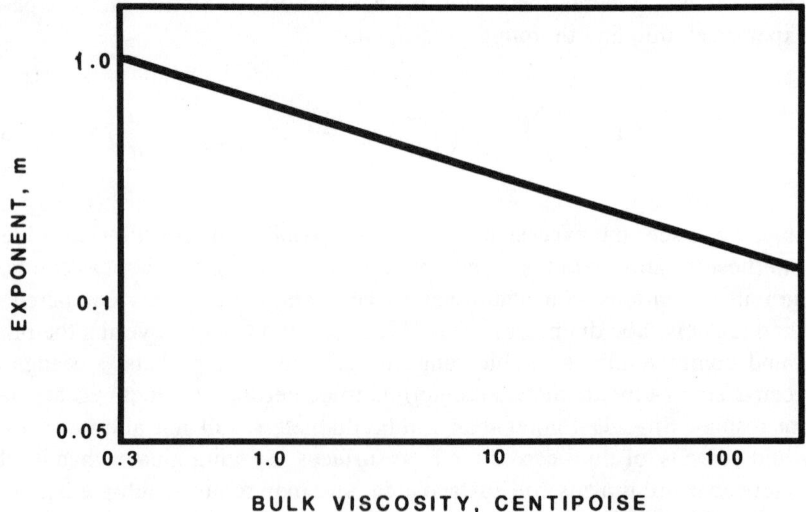

FIGURE 2-30. Exponent m versus bulk viscosity. (*Mixing Equipment, Inc.*)

Vertical Plate or Panel Coil

Plate or panel coils are variations of vertical tubes and are easier to install and maintain. The baffling effect with plate or panelcoils is greater than with vertical tubes. However, they restrict fluid flow more around the heat transfer surface and hence spacing is critical.

The following correlations have been proposed for heat transfer with vertical plate coils. For $N_{Re} < 1400$:

$$\frac{h_o L}{K} = 0.1788 \left(\frac{D^2 N\rho}{\mu}\right)^{0.448} \left(\frac{C_p\mu}{K}\right)^{1/3} \left(\frac{\mu}{\mu_f}\right)^{1/2} \tag{40}$$

For $N_{Re} > 4000$

$$\frac{h_o L}{K} = 0.0317 \left(\frac{D^2 N\rho}{\mu}\right)^{2/3} \left(\frac{C_p\mu}{K}\right)^{1/3} \left(\frac{\mu}{\mu_f}\right)^{1/2} \tag{41}$$

where: L = plate width per plate or panel coil
μ_f = viscosity of fluid at mean film temperature along the platecoil surface.

The angle at which the plate coil intersects the tank radius has been found to have no significant effect on heat transfer or power consumption. This means that plate coils can be angled for maximum heat transfer surface area within a tank. However, 90° is best to minimize stagnant flow.

Panel coils should be equally spaced around the inside tank wall, and they should be spaced no closer than 10° apart from each other to allow fluid to flow between them.

Overcrowding the inside of tanks with panel coils, helical coils, or vertical tubes spaced too close together can result in the agitator performing in essence in an unbaffled tank. Tight spacing of heat transfer surface will not provide flow through and around the surface and the flow pattern will take on a swirling regime. Inadequate velocities across the surfaces will not achieve the anticipated mixer-side heat transfer film coefficient.

Theoretically, the best heat transfer coefficient for the least mixer horsepower can be obtained if swirling due to lack of baffles is allowed. However, unbaffled systems are limited to relatively narrow ranges of Reynolds number because of the limiting condition of totally rotational motion. Since baffled systems are capable of much higher Reynolds numbers, they are ultimately capable of higher heat transfer coefficients.

Although agitation is necessary to provide the driving force for good forced-convection heat transfer, sizing an agitator to achieve a desired heat transfer coefficient may be impractical. As a typical case, Eq. (37) can be expanded to show the effects of each variable on the heat-transfer coefficient.

$$h_o \propto \mu^{-0.30} D^{1.44} N^{0.67} T^{-0.6} K^{0.63} d^{-0.5} \rho^{0.67} C_p^{0.37} \left(\frac{\mu}{\mu_s}\right)^M \quad (42)$$

This relationship is very useful with an existing installation when a single change is being considered (i.e. retrofitting mixer impellers, etc.), and for new installations to help make comparative evaluations of the variables. The correlation is specific to helical coil heat transfer systems, but can be used generally.

Combining Eqs. (37) with the power number relationship ($N_p = \text{Hp}/\rho N^3 D^5$), the heat transfer coefficient can be related to horsepower directly for comparison of a particular impeller type. At constant impeller diameter D,

$$h_o \propto (\text{Hp})^{0.22} \quad (43)$$

For a mixer being designed, where the impeller diameter has not been fixed, the effect on the heat transfer coefficient of increasing horsepower at, for example, constant speed:

$$h_o \propto (\text{Hp})^{0.29} \quad (44)$$

In order to double the mixer-side heat transfer coefficient, the horsepower would have to be increased by a factor of 11–23, depending on what aspect of the mixer design could be altered. Thus, as long as the fluid regime is one of forced convection circulation, increasing the level of mixing through greater power consumption will not economically provide a significant change in the heat transfer coefficient. Changes in heat transfer coefficient will be more effective if made by a change in factors other than mixing variables (i.e., type of heat transfer surface, heating or cooling medium, etc.)

Many of the variables in Eq. (42) deal with properties of the fluid and heat transfer surface dimensions. Although they are not affected by the mixer, the heat transfer coefficient may be altered significantly if they are changed.

In many applications other mixing requirements such as dissolving, solids suspension or blending will govern the ultimate design of the agitator. These operations must always be accounted for and the mixer designed for each specific application to determine which one is the controlling factor. When heat transfer is required in addition to another application, say solids suspension for example, care must be taken to design the mixer such that the solids suspension requirement is met. A reasonable approach is to select an agitator that will ensure forced-convection circulation, provide bulk-fluid motion, and meet the other mixing requirements, and afterwards change other process variables, such as temperature, temperature differential driving force, surface area, etc. to accommodate the heat transfer requirement.

The energy consumption of a mixing impeller is transformed into heat input into the tank contents. An input of 2550 Btu per hour additional heat load is transferred into the batch for each mixer Hp per hour. In moderately viscous fluid mixing application, or those applications requiring high power levels, the heat load due to the mixing impeller may be a significant factor. For example, the temperature rise in a batch due to mixing alone will typically be 0.2–0.3°F at a power level of 1 Hp per 1,000 gallons of fluid.

In general, test results have shown that approximately the same heat transfer film coefficient is obtained regardless of whether the impeller is a flat blade radial turbine, a pitch-bladed axial turbine, a flow efficient hydrofoil or propeller. Once in a forced-convection fluid regime, in which temperature gradients are low, the direction of primary pumping makes little difference on the heat transfer coefficient. Thus the impeller type should be selected based on other application criteria or to maximize flow while minimizing power to optimize the mixer side heat transfer coefficient.

Usually axial flow turbines are best suited for heat transfer applications, as they are for any other flow controlled mixing requirement. They produce the greatest amount of flow per unit horsepower with reasonable capital cost implications. Along this line of thought, a hydrofoil impeller is the optimum choice of impeller geometry, since it affords the highest flow efficiency of the family of axial flow turbines. In the case of smaller batches with portable mixers, propellers will be an economical choice of impeller. When space is a problem, a higher power number flat bladed radial turbine is recommended. In some cases, batch depth may predicate that multiple impellers be installed to ensure forced convection throughout the tank contents.

Impellers should be located approximately at mid-depth. The mixer side heat transfer film coefficient will decrease as the impeller is moved closer to the bottom of the tank. Mixing impellers located near the tank bottom increase local turbulence, at the expense however of diminished velocity turbulence at the heat transfer surfaces located far from the turbine. For this reason, the mixer-side film coefficient decreases with decreasing impeller off-bottom elevations. Typically, 20% better heat transfer rates have been reported with central impeller location compared to near-bottom location. Other factors, however, such as simultaneous solids suspension requirement and the necessity to accommodate lower batch levels, may preclude locating the impeller at the optimum location for heat transfer.

In general, hydrofoils optimize heat transfer rates because these impellers produce the greatest flow per unit power. A further benefit may be accrued because hydrofoils can be located at mid-depth, while still producing and maintaining high flow per unit power.

Convective heat transfer involves the transfer of heat due to both bulk transport and macroscopic mixing of elements of the fluids. Forced convection heat transfer is the result of fluid motion due to a mixer.

Figure 2-31 is a typical plot showing the effect of agitator speed on measured heat transfer coefficient. In the middle speed range is a zone where the physics of heat transfer by forced convection predominate. In this range there is a straight-line relationship with a slope of 0.67, which is the exponent on Reynolds number in the heat transfer correlation.

Below point A, fluid turbulence decreases to where natural convection forces begin to influence heat transfer. At an agitator speed of zero, natural convection alone provides heat transfer, and the two curves converge.

At agitator speeds above point B, fluid motion is so turbulent that further increases in such motion provides little driving force for additional heat transfer. At these speeds, the surface motion is likely to be so intense that air will be entrained into the liquid, further complicating heat transfer relationships. Most well baffled mixing systems operate below point B.

The greatest incremental gains in heat transfer efficiency are achieved by installing an agitator in a tank previously limited to natural convection heat transfer. For example, an agitator operated at 80 rpm (Fig. 2-31) increased the

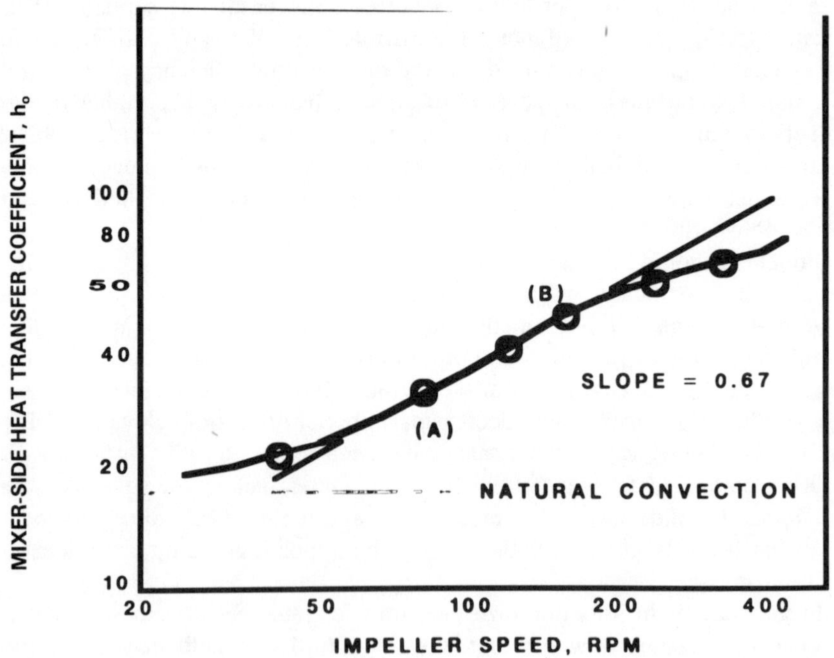

FIGURE 2-31. Typical logarithmic plot of heat transfer coefficient versus impeller speed. (*Mixing Equipment, Inc.*)

mixer side heat transfer film coefficient by nearly 100% over the coefficient at 0 rpm. Yet to double the heat transfer coefficient again would require a speed increase of over 300% (or over 27 times as much power), due to diminishing returns for heat transfer with power in this speed range.

It is imperative to operate in the forced convection zone of Figure 2-31. Experimental data may be plotted to generate the curve or if such data is not available, a fast selection may be made without excessive cost by using twice the minimum power required for continuous blending. This is also a good design method to follow in making preliminary budget estimates of mixer size and cost for heat transfer applications.

Empirical relationships have been developed to determine minimum agitator speed to ensure forced convection. For example, the following relationship has been presented to be used with Eq. (39):

$$N \geq 43 \left(\frac{T}{D}\right)^{2.5} \left(\frac{B}{2}\right)^{0.3} (N\nu)^{0.2} \tag{45}$$

where: N = impeller speed, rpm
T = tank diameter
D = impeller diameter
B = baffle width
ν = kinematic viscosity

As long as the inequality is satisfied and the mixing is not too severe, Eq. (39) defines the heat transfer characteristics of the system tested. However, such empirical relationships tend to be particular to the system studies. Visual observation is usually the only test necessary to confirm if a mixing system is one of forced circulation.

Preliminary estimates of mixer-side heat transfer coefficients may be made by using Fig. 2-32. Typical mixer-side coefficients are shown for cooling and heating of organic liquids. This will form the basis of an order of magnitude estimate of coefficients suitable for preliminary design work. The coefficients will obviously change, depending upon the system and characteristics of the system, and the mixer selection. The values shown in Fig. 2-32 have been generated for vertical tubes or a single bank of helical coils. For properly baffled jacketed tanks, the coefficients should be taken as 65% of the values obtained by using Fig. 2-32.

Because of their higher thermal conductivities, the coefficient of aqueous solutions of inorganics will be three to four times greater than those for organic liquids in Fig. 2-32.

Control of heat flow in a desired manner is a most important consideration in chemical engineering. Most chemical and biological processes are either endothermic or exothermic (heat transfer into or removed from the tank contents

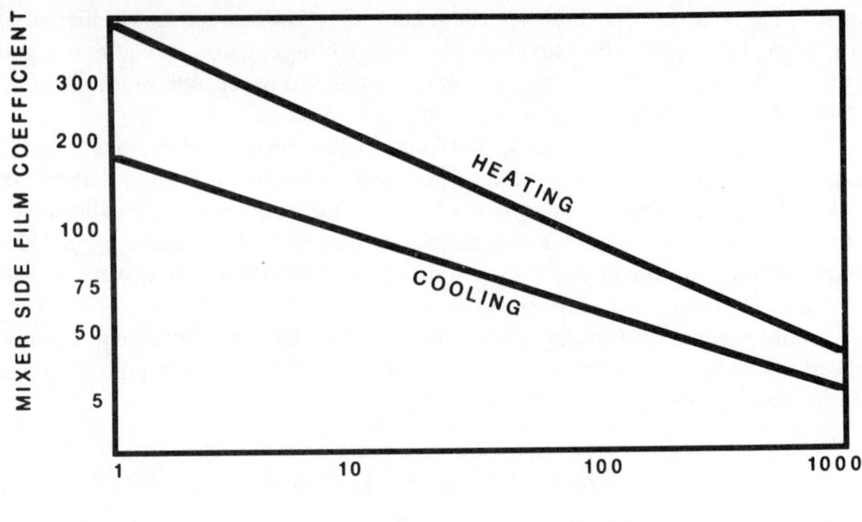

FIGURE 2-32. Typical mixer-side film coefficients for heating and cooling organic chemicals. (Thermal effectiveness 1.54.) (*Mixing Equipment, Inc.*)

respectively). About 80% of all chemical processing mixer applications involve heat transfer of some degree.

Typical systems in which heat transfer is an important process parameter are:

1. Chemical reactions
2. Polymerization
3. Fermentation
4. Nitration
5. Hydrogenations

Mixing can only affect the mixer side heat transfer film coefficient h_o, which represents only a part of the overall heat transfer equation. In order to determine when the degree of mixing intensity will produce a significant change in h_o, a study of the fluid regime is required.

If the fluid regime is forced convection (i.e., the fluid being heated or cooled is in turbulent flow at the heat transfer surface), increasing the mixing intensity (power) will not economically provide a significant change in h_o due to diminishing return of flow velocity on power.

$$h_o \propto \text{Hp}^{0.22} \tag{46}$$

For example, increasing horsepower by a factor of 10 would only increase h_o by 66%, or a factor of 1.66.

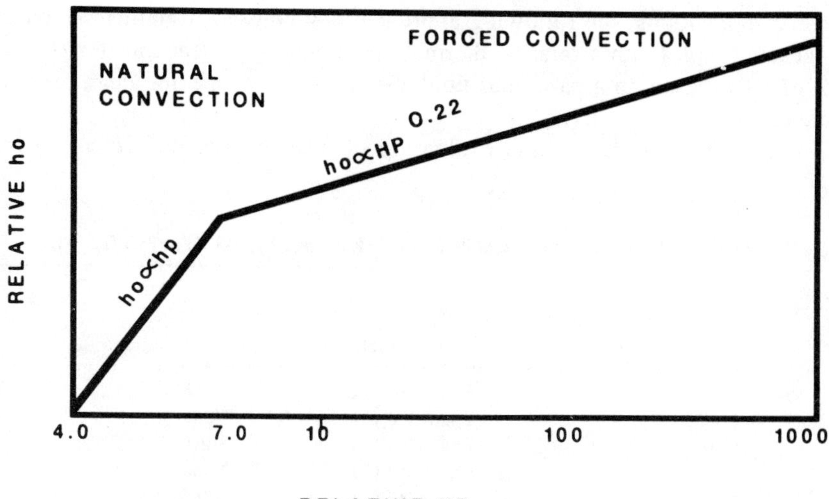

FIGURE 2-33. h_0 versus Hp (for determining heat transfer condition). (*Mixing Equipment, Inc.*)

If the fluid regime is laminar flow, the heat transfer condition is natural convection. This regime requires laminar flow of the tank contents across the heat transfer surface during heating or cooling cycles. Increasing the mixer intensity or power level produces a proportional increase in mixer-side film coefficient h_o:

$$h_o \propto \text{Hp} \tag{47}$$

In this scenario, delivering twice the power level will double the heat transfer coefficient, as long as laminar flow conditions are still maintained.

If horsepower is changed in a given mixing system and the corresponding value of h_o are calculated, a plot of h_o versus Hp will determine if the regime is natural or forced convection. Fig. 2-33 depicts this relationship.

To simplify the procedure in evaluating different agitator designs and alternative heat transfer surfaces to select the appropriate mixer for various heat

TABLE 2-6. Thermal Effectiveness of Heat Transfer Surfaces.

Tank jacket	1.0
Vertical tubes	1.54
Helical coils (1st bank)	1.54
Helical coils (2nd bank)	1.31
Helical coils (3rd bank)	1.08
Plate or panel coils	1.12

transfer applications, the following approach may be used. Calculate h_o based on jacketed tank (with a relative thermal effectiveness of heat transfer surface area of 1.0), based on a particular fluid specific gravity, specific heat, thermal conductivity and viscosity.

For relative comparison and evaluation of heat transfer coefficients using

TABLE 2-7. Jacketed Vessels, Overall Coefficients (*U* Expressed in BTU/hr/sq ft/°F).

Fluid inside Jacket	Fluid in Vessel	Wall Material	Agitation	*U*
Steam	Water	Enameled C.I.*	0–400 R.P.M.	96–120
Steam	Milk	Enameled C.I.	None	200
Steam	Milk	Enameled C.I.	Stirring	300
Steam	Milk boiling	Enameled C.I.	None	500
Steam	Milk	Enameled C.I.	200 R.P.M.	86
Steam	Fruit slurry	Enameled C.I.	None	33–90
Steam	Fruit slurry	Enameled C.I.	Stirring	154
Steam	Water	C.I. and loose lead lining	Agitated	4–9
Steam	Water	C.I. and loose lead lining	None	3
Steam	Boiling SO_2	Steel	None	60
Steam	Boiling water	Steel	None	187
Hot water	Warm water	Enameled C.I.	None	70
Cold water	Cold water	Enameled C.I.	None	43
Ice water	Cold water	Stoneware	Agitated	7
Ice water	Cold water	Stoneware	None	5
Brine, low velocity	Nitration slurry	—	35–58 R.P.M.	32–60
Water	Sodium alcoholate solution	"Frederking" (cast-incoil)	Agitated, baffled	80
Steam	Evaporating water	Copper	—	381
Steam	Evaporating water	Enamelware	—	36.7
Steam	Water	Copper	None	148
Steam	Water	Copper	Simple stirring	244
Steam	Boiling water	Copper	None	250
Steam	Paraffin wax	Copper	None	27.4
Steam	Paraffin wax	Cast iron	Scraper	107
Water	Paraffin wax	Copper	None	24.4
Water	Paraffin wax	Cast iron	Scraper	72.3
Steam	Solution	Cast iron	Double scrapers	175–210
Steam	Slurry	Cast iron	Double scrapers	160–175
Steam	Paste	Cast iron	Double scrapers	125–150
Steam	Lumpy mass	Cast iron	Double scrapers	75–96
Steam	Powder (5% moisture)	Cast iron	Double scrapers	41–51

*C.I. = cast iron.

different fluids, alternative surfaces, etc., the following corrections must be applied:

1. For adding vertical tubes or helical coils to a jacketed tank, correct h_o by:

$$\frac{h_{o_2}}{h_{o_1}} = 0.2\, d^{-0.5} T^{0.5} \qquad \left(\text{provided } \frac{d}{T} \leq 0.04\right) \tag{48}$$

2. Use the values of thermal effectiveness for various surfaces as a multiplier on h_o. (Table 2-6) The thermal effectiveness factor reflects on how well fluid in turbulent flow moves over and around a particular surface, utilizing as much contact area as possible with the surface.
3. For fluid specific heat or thermal conductivity factors, consider:

$$\frac{h_{o\,\text{fluid}}}{h_{o\,\text{water}}} = \left(\frac{C_{p\,\text{fluid}}}{C_{p\,\text{water}}}\right)^{0.37} \left(\frac{K_{\text{fluid}}}{K_{\text{water}}}\right)^{0.63} \tag{49}$$

4. If $\mu_s/\mu \neq 1.0$, use

$$h_o \propto \left(\frac{\mu_s}{\mu}\right)^{-m} \tag{50}$$

as a multiplier. The exponent m is obtained from Fig. 2-30.
5. Apply multipliers from steps 2, 3, and 4 to h_o used as a reference to obtain corrected values.

Various overall heat transfer coefficients, μ_o, for agitated jacketed tanks are shown in the Table 2-7.

J. CRYSTALLIZATION

Crystallization is the reverse of dissolving and is accomplished either by cooling a saturated solution to precipitate out crystals, or by heating a solution to drive off the solvent. The agitator selection usually resolves itself into (1) a heat transfer application necessitating good velocity and flow of the solution across the heat transfer surface, (2) suspension of the crystals being formed, and (3) satisfactory handling of the crystals after formation.

The problem of handling crystals usually dictates the choice of impeller based on the proper balance of flow and shear development. Some crystals are shear sensitive and friable in nature, likely to fracture in high shear environments. These crystals must be agitated gently with hydrofoil turbines to minimize shear.

Other crystals may tend to form on the cooling surfaces, actually fouling the heat transfer surface and diminishing the heat transfer rate. These crystals must be scraped off the heat transfer surface (by the impeller preferably) to renew a clean surface area and maximize heat transfer. Anchor type impellers running at close clearance to the jacketed tank wall can be equipped with scrapers to wipe the inside wall surfaces clean on a continuous operation basis. These scrapers can be teflon spring-loaded extensions of the impeller blades. Teflon has excellent abrasion and corrosion resistant properties, is inert to many materials in contact with it, and can be fabricated into blade extensions, making it an excellent material to scrape the tank walls. By spring loading these extensions, one can be assured of continuous teflon to tank wall contact, even as the teflon wears over continuous operation and rubbing action.

Some crystallizers in continuous flow may be designed to concentrate crystals at the bottom of the tank for drawoff. Proper agitator selection requires knowing the crystal particle size distribution, specific gravity, viscosity of solution, and concentration of crystals in solution. Other applications may require that the crystals be maintained uniformly in suspension. In other crystallizers, the solids may deposit out to the extent that circulation and motion is impeded because of the high concentration of solids in suspension.

Crystallizers fall into three basic types. The first is a conventional vessel with either jacket or internal coils or both for heat transfer. The open impeller is usually a low speed, high flow design providing good volumetric flow rates at moderate to low velocity. This type of crystallizer is useful for shear sensitive, easily damaged crystals, or where solids content builds up to a high level.

The second is a conventional vessel fitted with a draft tube in addition to a jacket or coil. The draft tube circulator provides positive flow control by preventing short circuiting of the incoming solution out the discharge without proper retention and circulation. Usually high speed impellers such as propellers and hydrofoils are used. This type of crystallizer is useful where crystal size is not important and for crystals that are not shear sensitive. The draft tube circulator crystallizer imparts greater shear rates on crystals than open impeller crystallizers because: (1) The crystals impact at higher localized velocity on the tank floor, (2) high velocity through the draft tube, (3) significant transitional velocities at the inlet and outlet of the tube where the flow turns from vertical to horizontal and vice versa, (4) close clearance running of impeller in the draft tube, and (5) high speed operation of the draft tube circulator mixer.

Draft tube circulator crystallizers can pump up or down in the draft tube. Up-pumping reduces overall shear on the crystals being formed since the crystals do not impact on the tank bottom. However, care must be exercised to do the following with up-pumping draft tube circulator crystallizers: (a) Locate feed inlets on the suction side of the circulator impeller, but not directed onto the impeller itself—i.e., near the tube bottom directed at the bottom end of the draft tube. (b) Locate the impeller close to the bottom of the tube. (c) Provide smooth transitions of flow into the bottom of the draft tube with a beveled tube inlet section and teardrop transition along the bottom of the tank to help direct flow along the tank floor into the bottom of the tube. (d) Locate the draft tube at proper clearance to the tank bottom and side walls. (e) Provide a truncated cone and dished bottom transition from a traditional vertical cylindrical vessel, etc.

A third type of crystallizer is one used with "pusher type" impellers such as helical ribbons, gates, or anchor turbines running in close clearance or actually scraping the vessel walls. This configuration is used for high viscosity crystallizations and situations where crystals form on the heat transfer surface and require wiping to avoid fouling.

Additional data required for crystallization agitation requirements are as follows:

1. General process description, with emphasis on progressive changes in crystal structure and slurry characteristics as crystallization proceeds
2. Shear sensitivity of crystals
3. Fouling characteristics of crystals on surfaces
4. Operating temperature and pressure of the crystallizer
5. Description of heat transfer surfaces
6. The nature of the crystals, whether they should be maintained uniformly in suspension or permitted to concentrate in the bottom of the crystallizer with off-bottom suspension and bottom discharge
7. Specific gravities of the crystals and the mother liquor, particle size distribution of the crystals, viscosity of the liquor, settling velocity of the crystals
8. Crystal formation rates and method of formation (i.e., do they form as slurry or do they deposit out on heat transfer surfaces?)
9. Latent heat of crystallization and sensible heat transfer for the solution in question to provide overall heat transfer requirement

For example, a 15,000 gallon crystallizer is equipped with a central draft tube and used for forming solids at 25% by weight maximum with settling velocity up to 5 feet per minute. Crystals will be continuously drawn off the tank bottom. Here on-bottom motion of the crystals is desired, since a greater concentration of solids at the drawoff point is desired. A 10 Hp axial flow impeller to provide on-bottom motion of crystals would suffice. The design of the mixer is predicated on providing on-bottom motion of solids with a settling velocity of 5 feet per minute maximum.

3

Shear Controlled Mixing Applications

A. GAS–LIQUID DISPERSIONS AND MASS TRANSFER

There are two distinct areas of gas dispersion:

1. Physical dispersion
2. Dispersion for mass transfer

Radial flow impellers and wide blade, high solidity axial flow hydrofoils are used for gas dispersion mixing because they have lower flow to head ratios (and consequently greater head to flow balance) than other impellers, and head manifests itself in shear. For gas dispersion applications, not only is flow important, but also shear development by the mixing turbine. Flat bladed radial flow turbines and wide blade axial flow hydrofoils provide the proper balance of flow and shear required in gas dispersion applications. Fig. 3-1 shows that these impellers are best suited for dispersing gas in a liquid.

Flow is essential in gas–liquid dispersion requirements because it provides the driving force (reflected in bulk fluid velocity) for shear forces and physical dispersion of the gas. Flow also provides gas–liquid contacting when the solubility of gas in the liquid is great enough that the controlling step is physical dispersion and not mass transfer of the gas into the liquid phase.

A mixer is required in a gas–liquid system to:

1. Increase the interfacial area of gas bubbles by shearing the gas into small bubbles
2. Maximize the residence or retention time of gas in the liquid
3. Provide bulk fluid velocity significant enough to drive the bubbles of gas to the tank bottom

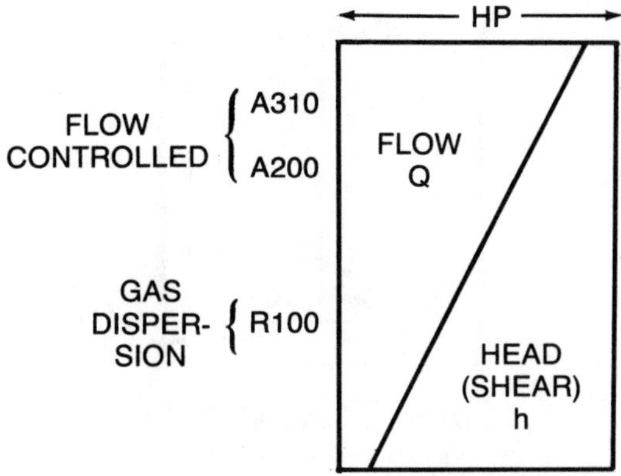

FIGURE 3-1. Flow–shear spectrum for mixing applications. (*Mixing Equipment, Inc.*)

Shear is the gradient of flow velocity across the mixing impeller blade. This differential in velocity profile must be adequate in gas–liquid applications to produce smaller gas bubbles from larger ones to increase surface contact area between gas and liquid. This will inherently decrease gas–liquid reaction time while simultaneously increasing system efficiency due to increased gas holdup time. Smaller gas bubbles rise more slowly than larger ones. Increased gas holdup time means a longer residence time and, therefore, higher efficiency when adequate shear is applied to the fluid.

B. PHYSICAL DISPERSION

In order for gas to be dispersed by a mixer, the applied impeller horsepower (shaft horsepower at the impeller) must be equal to or greater than the isothermal gas expansion horsepower. Otherwise the gas will overcome the driving force of the mixer bypassing the impeller, without being dispersed. As shaft power is increased, physical dispersion patterns of gas in the liquid are intensified. Without adequate shaft horsepower to overcome the gas expansion horsepower, "geysering" would occur. This type of flow pattern is shown in Fig. 3-2.

When shaft horsepower at the mixing impeller is greater than gas expansion power, gas dispersion will occur and the *degree* of dispersion will be a function of the applied shaft horsepower. The basic three degrees of physical gas dispersion are:

1. Minimum
2. Intimate
3. Uniform

Fig. 3-3 through 3-5 illustrate these basic degrees of gas dispersion.

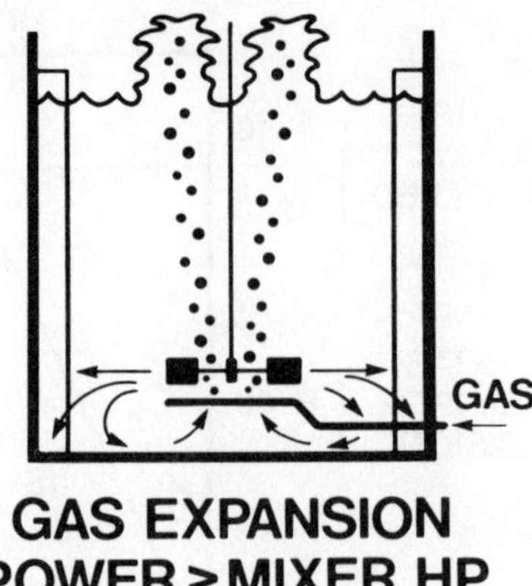

FIGURE 3-2. "Geysering" flow pattern. (*Mixing Equipment, Inc.*)

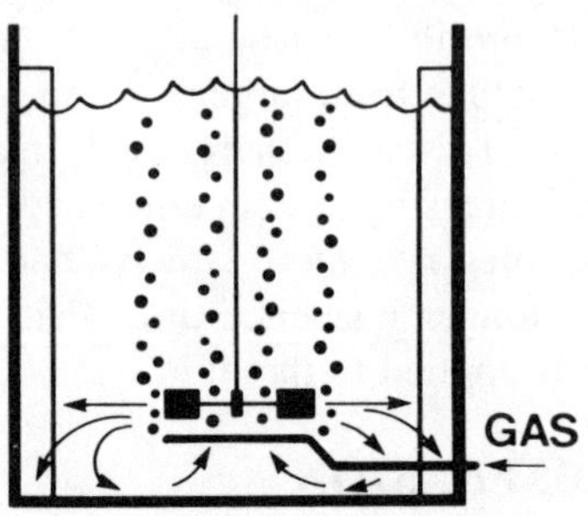

FIGURE 3-3. Minimum dispersion. (*Mixing Equipment, Inc.*)

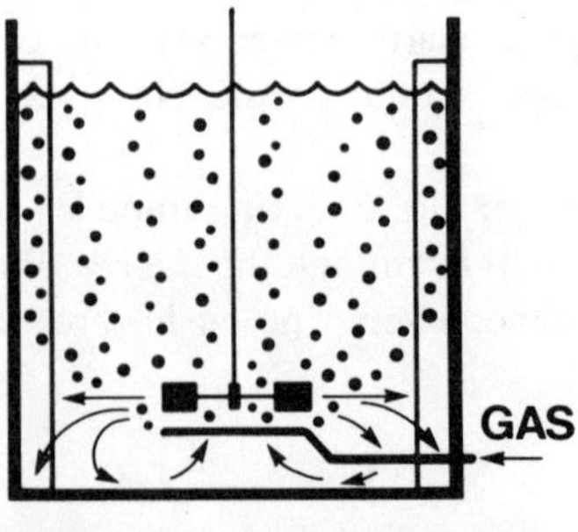

FIGURE 3-4. Intimate dispersion. (*Mixing Equipment, Inc.*)

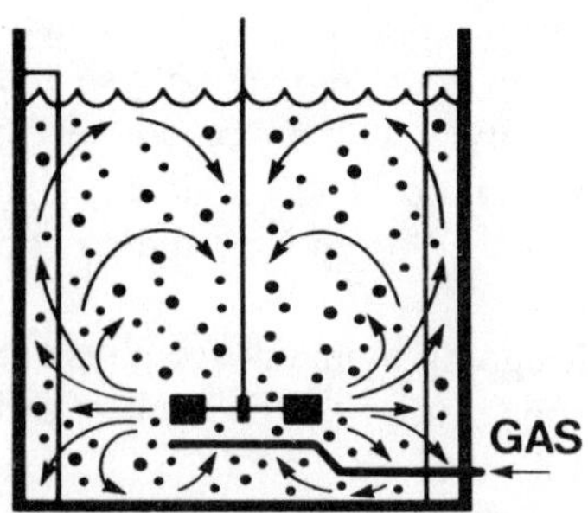

FIGURE 3-5. Uniform dispersion. (*Mixing Equipment, Inc.*)

In order to achieve the proper levels of dispersion for the basic degrees of physical gas dispersion, the following shaft power levels at the turbine are required:

1. For minimum gas dispersion, 1.0 times gas expansion Hp
2. For intimate gas dispersion, from 1.0 to 3.0 times gas expansion Hp
3. For uniform gas dispersion, 3.0 times gas expansion Hp

To calculate the isothermal gas expansion horsepower, one must start with the general equation for work, since work is what is accomplished by the gas:

$$W = P_1 V_1 \ln (P_1/P_2) \tag{1}$$

$$\text{Hp/Vol} = \frac{(4.051 \times 10^3) \cdot (P_2 + 62.4Z\rho)}{Z} \ln \left(\frac{P_2 + 62.4\,Z\rho}{P_2}\right) F_1 \tag{2}$$

where: W = work performed, ft-lb/sec
P_1 = pressure at tank bottom, lb/ft^2
V_1 = volumetric flow rate of gas, ft^3/sec
P_2 = pressure at tank top, lb/ft^2
Z = batch height or depth, ft
Vol = batch volume (ungassed), thousands of gallons
ρ = ungassed liquid specific gravity
F_1 = superficial gas velocity at gas inlet, ft/minute

$$F_1 = Q_1/A \tag{3}$$

where: Q_1 = volumetric gas flow rate at the gas inlet location, actual cubic feet per minute (ACFM)
A = cross sectional area of tank, ft.

Once the gas expansion horsepower is calculated, the agitator can be designed to accomplish the various degree of physical dispersion. However, mixer horsepower is not the only important parameter for the required dispersion. Impeller to tank diameter ratio (*D/T*) is also a significant variable. The most effective ratios to use for the horsepower and gas rates required are depicted in the matrix in Fig. 3-6.

It may seem that designing agitators for physical gas dispersion is the only criterion for consideration. However, a check must be made to be sure that the mixing impeller is not flooded by the gas.

In a gas–liquid mixing system, both the mixer and gas influence the flow pattern. The mixer effects flow patterns because of its head/flow development. Gas affects the flow pattern because of bubble formation and rise velocity. When the gas is the controlling factor, the mixer loses control of the flow pattern and the system flow pattern becomes gas controlled. The result is loss of dispersion, poor mixing, and loss of gas holdup and retention time. In essence a "geysering" flow pattern similar to that shown in Fig. 3-2 takes over. The impeller is therefore said to be flooded by the gas.

When the mixer is controlling the flow pattern, gas is dispersed in such a way that one of the patterns depicted in Figs. 3-3 through 3-5 develops. Gas is dispersed and holdup of gas in the liquid phase can actually be measured (by measuring the depth of liquid expansion under gassed conditions and calculating the fluid volume increase).

There is a maximum gas rate that a specific impeller type, of certain diameter and running speed, can disperse without being flooded. If this gas rate is exceeded, the impeller will become flooded and the system flow pattern will be gas controlled, creating a "geysering" effect. Some impellers such as axial flow flat bladed turbines, narrow bladed axial flow hydrofoils, and propellers can disperse gas at only very low rates without flooding and as such are impractical for consideration in gas dispersion applications at even modest rates of gas introduction. Radial flow flat bladed turbines and wide blade axial flow hydro-

MOST EFFECTIVE D/T RATIOS		
HP / GAS RATE	HIGH	LOW
HIGH	0.17	0.33 – 0.50
LOW	0.17 – 0.50	0.17

FIGURE 3-6. Optimum *D/T* ratio for gas dispersion. (*Mixing Equipment, Inc.*)

foils can disperse modest to high gas rates without flooding and are considered the ideal impeller geometries for gas dispersion mixing requirements. All impellers regardless of geometry, diameter, and/or speed will have a threshold of gas rate before flooding occurs.

C. MIXER POWER LOADING IN GAS DISPERSION APPLICATIONS

The question to be asked is, should the mixer be designed for proper power loading in the gassed or ungassed state?

Power consumption of a flat-bladed radial flow turbine (with a disk and blades) in a gas–liquid system is less than the power consumed when the gas is off. In the gassed scenario the impeller "sees" and operates in a region of a gas–liquid mixture, with lower specific gravity than that of the ungassed liquid. The ratio of gassed to ungassed power draw, P_g/P_0, is called the *K* factor of the impeller. It is a function of the type and geometry of impeller (wide bladed axial flow hydrofoils have higher *K* factors, even approaching unity under certain conditions, than radial flow disk turbines with flat blades at similar gas rate and diameter). Other variables affecting *K* factors are gas rate and impeller speed. Fig. 3-7 shows the relationships for a radial flow turbine with disk and flat blades in an air–water gassed system,

where: P_g = gassed shaft horsepower
P_0 = ungassed shaft horsepower
Q = air flow rate, actual cubic feet per minute (ACFM)
$N_Q = Q/ND^3$, dimensionless
N = impeller speed, rpm
D = impeller diameter, ft

Assume that for a given level of gas dispersion that P_g = 13.5 Hp and P_0 = 27 Hp for a *K* factor = 0.5. If the mixer is sized on the ungassed basis, a larger mixer would be required than one designed on a gassed basis. However, we would be assured of operating safely, without concern for power overload, in either gassed or ungassed scenarios.

On the other hand, if we sized the agitator on the basis of gassed power draw, an overload condition might exist in a reduced gas flow or most certainly in an ungassed situation. Does this mean that the agitator is limited to operate only in the gassed condition? The answer is "yes" with a single speed motor and "no" with a two speed motor. If a two speed motor is supplied, both modes of operation (i.e., gassed and ungassed) could be pursued without oversizing the agitator, while still satisfying the process results. For an agitator with a

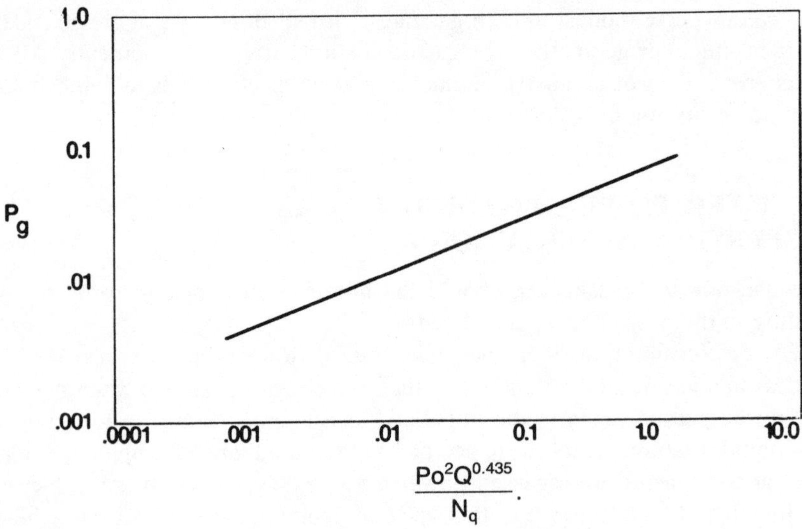

FIGURE 3-7. Relationship of *K* factor and gas flow for air–water agitated systems. (*Mixing Equipment, Inc.*)

design premise of power draw based on operating in a fully gassed regime, the higher motor speed (and hence greater mixer impeller speed) can be chosen. Upon reduction of gas rate, including complete shutting off of the gas, the motor speed can be switched to the low speed to compensate for the higher power draw at reduced gas rate, without jeopardizing an overload of power on the agitator. The change in motor speed can be accommodated automatically by the use of a gas flow switch interlocked with the motor to reduce speed based on reduced (or complete loss of) gas flow rate. Infinitely adjustable motor speeds utilizing frequency controllers and/or hydraulic motors can accommodate subtle changes in gas rate by subtle adjustments to motor speed. Loading a mixer based on the gassed horsepower draw and using a two speed, variable speed motor or gas flow–motor speed interlock offers these advantages:

1. Smaller mixer size due to lower power and torque requirement, therefore lower initial purchase and installation cost
2. Lower operating cost due to reduced power draw during reduced gas conditions
3. Reduced spare parts cost (and faster delivery of smaller sized mixer components)
4. Easier and less expensive maintenance cost
5. Smaller, less expensive starter box

D. SPARGING DEVICES

Gas introduction into the liquid may be accomplished in many ways. The manner in which gas is introduced is not arbitrary, however. Three common sparge designs are illustrated in Fig. 3-8, 3-9, and 3-10.

A sparge ring (Fig. 3-8) is best suited for clean gas and liquids and nonslurries. One must be careful to avoid plugging off the small sparge ring holes with dirt, fouling or settled solids from a slurry. Mass transfer is optimized with the sparge ring due to the gas discharging from the ring as uniformly distributed small diameter bubbles.

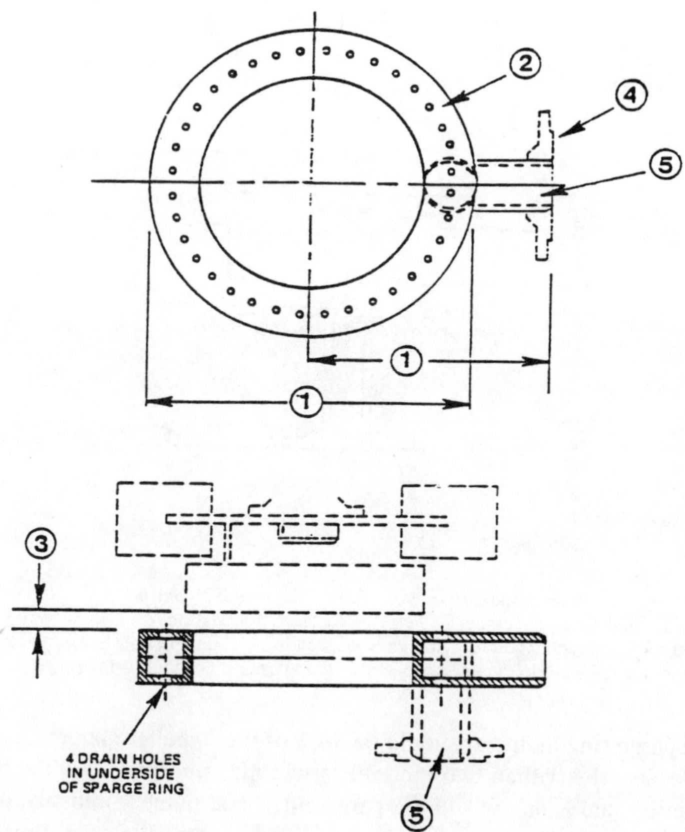

FIGURE 3-8. Gas inlet sparge ring design. (*Mixing Equipment, Inc.*) Notes: (1) Sparge ring diameter and inlet pipe extension selected for each application. (2) Gas outlet holes for each application. (3) Sparge location and clearance to impeller blades determined for each application. (4) Pipe flange connections are alternatives to all welded piping. (5) Pipe extensions can be horizontal or vertical downward.

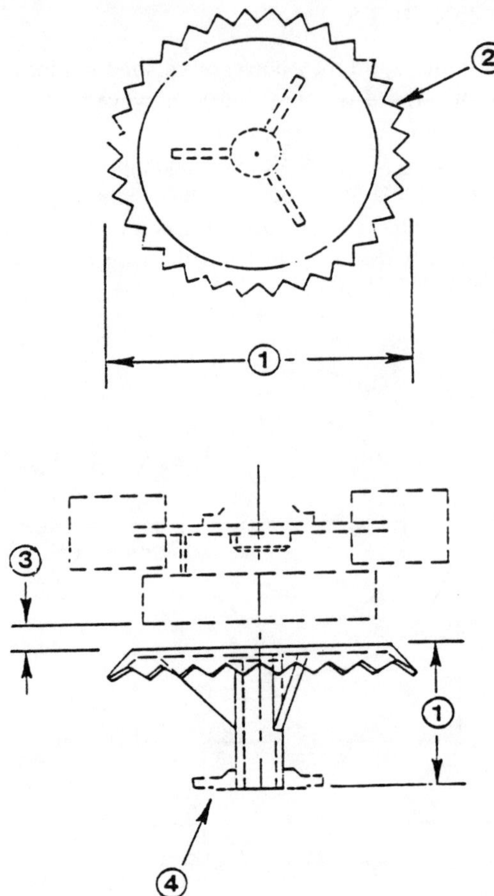

FIGURE 3-9. Gas inlet sparge plate assembly. (*Mixing Equipment, Inc.*) Notes: (1) Sparge plate diameter and inlet pipe extension selected for each application. (2) Gas dispersion serrations determined for each application. (3) Sparge plate location and impeller blade clearance determined for each application. (4) Pipe flange connections are alternatives to all welded piping.

The sparge ring diameter should be 80% of the impeller diameter and should be located no closer than two inches running clearance between the bottom of the impeller blade and the top of sparge ring. The number and size of the gas outlet holes are predicted on gas rate and allowable pressure drop. Pressure drop through the holes should be limited to 2–5 psi to pass the desired volumetric flow rate of gas, and avoid plugging of holes of small diameter.

Figure 3-11 shows the effect of sparge ring diameter on mass-transfer coefficient, all other things being equal. The optimum ratio of sparge ring to im-

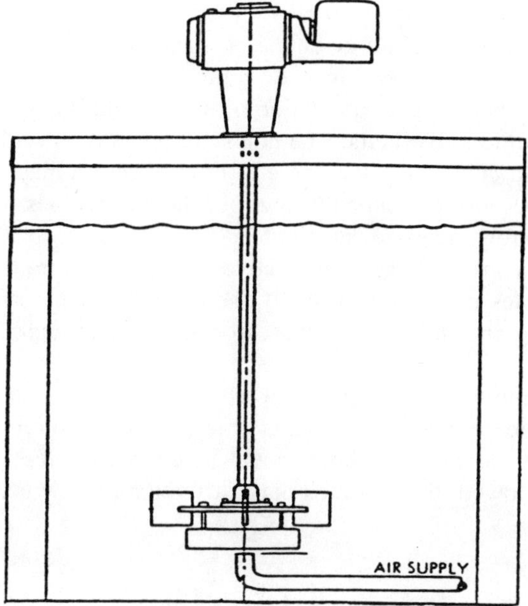

FIGURE 3-10. Open pipe sparging system. (*Mixing Equipment, Inc.*)

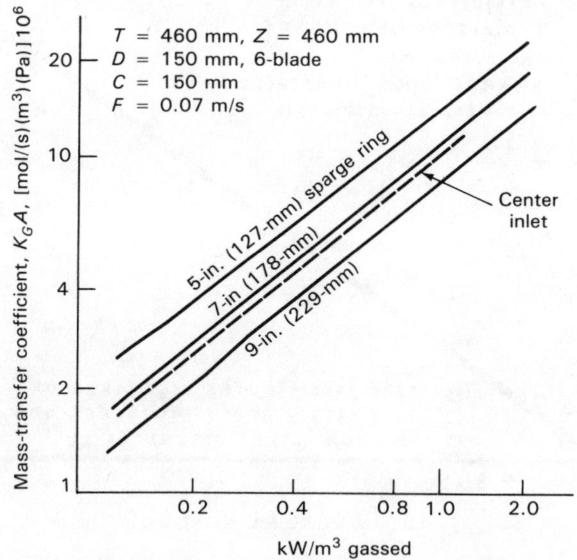

FIGURE 3-11. Coefficients for various sparge ring diameters. (*Mixing Equipment, Inc.*)

peller diameter is on the order of 0.8. This ratio can provide an increase of 10–15% in mass transfer rate, at a given gas rate and power level.

In the area where the mixer flow pattern prevails, there is very little effective difference due to the size and orientation of holes in the sparge ring. Fig. 3-12 shows that, regardless of whether the holes are pointed upward or downward, or are of any particular diameter, there is little effect on mass transfer coefficients. This is because the impeller, and not the gas bubbles produced by the sparge ring, controls the gas dispersion.

In designing a sparge ring, the objective is to get uniform flow distribution out of all the holes in the ring. Usually the diameter of the ring and pipe diameter are given and the design focuses on the number and diameter of the holes.

A rule-of-thumb for the design is that the ratios of kinetic energy in the inlet stream to pressure drop across the outlet hole and of friction loss in the ring to pressure drop across the outlet hole should be equal to or less than one-tenth. If this rule is followed, the expected maldistribution of flow is less than 5%.

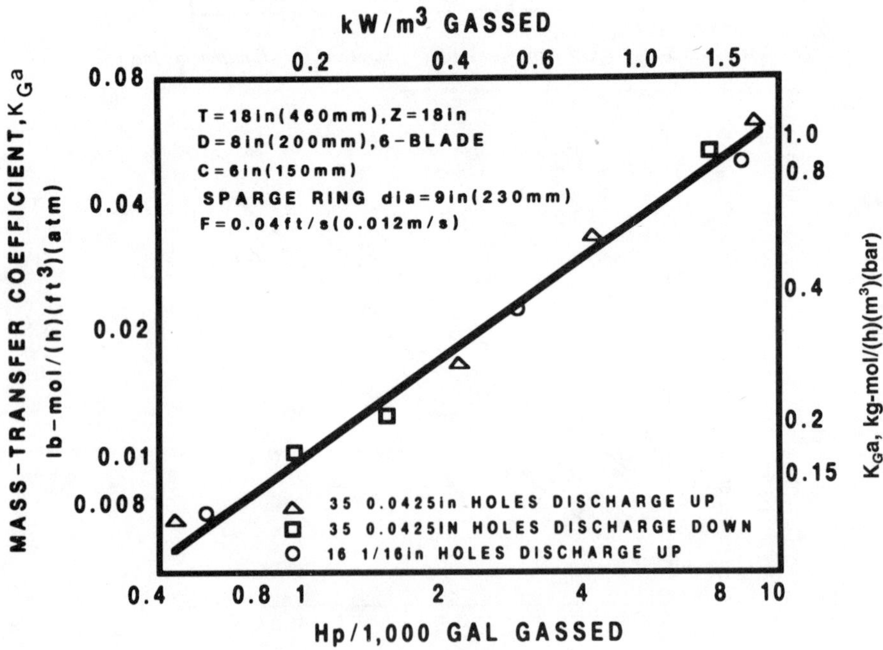

FIGURE 3-12. Coefficients with various hole diameters and numbers of holes. (*Mixing Equipment, Inc.*)

These constraints can be restated:

$$\frac{\alpha\rho\ \overline{V}_p^2/2}{\Delta P_0} \leq 0.1 \tag{4}$$

and

$$\frac{\Delta P_p}{\Delta P_0} \leq 0.1 \tag{5}$$

where: ρ = fluid density
$\overline{V}_p$ = average inlet pipe velocity
ΔP_0 = pressure drop across each hole
ΔP_p = pressure drop in ring caused by friction

The correction factor α is approximately 1.05 for turbulent flow.

To convert Eq. (4) to a useful form, we first calculate ΔP_0 using a relationship for turbulent flow through a sharp-edged orifice (with an A_0/A_p ratio between 0.05 and 0.70):

$$\Delta P_0 \sim 2.6\ \rho\ \frac{\overline{V}_0^{\,2}}{2}\ [1 - (A_0/A_p)^2] \tag{6}$$

where: $\overline{V}_0$ = average velocity of gas leaving each hole
A_0 = area of one hole
A_p = cross-sectional area of ring pipe

By continuity, one knows that $A_0\ \overline{V}_0\ N = A_p\ \overline{V}_p$, with N = number of holes.

We can convert Eq. (4)to an expression for hole or orifice diameter:

$$D_0 \leq D_P/(1 + 4.04\ N^2)^{1/4} \tag{7}$$

but since N is usually large, this further reduces to a simple constraint:

$$D_0 \leq 0.7\ D_p/\sqrt{N} \tag{8}$$

where: D_0 = diameter of each hole or orifice
D_p = diameter of ring pipe

To convert Eq. (5), we can estimate the friction loss in the ring pipe as approximately one velocity head per 150 pipe diameter:

$$\Delta P_p \approx \frac{\pi D_R}{150\ D_P} \rho \frac{\overline{V}_P^2}{2} \tag{9}$$

with D_R = diameter of sparge ring.

Combining Eq. (9) with Eq. (5) and (6), we get the other constraint:

$$D_0 \leq D_P \left(\frac{1 + \pi D_R\ N^2}{39 D_P} \right)^{1/4} \tag{10}$$

If $\pi\ D_R/D_P$ is less than 150, Eq. (8) is the only constraint that should be considered.

Drain holes should be drilled on the bottom side of the sparge ring to promote self cleaning and prevent plugging of the ring.

The sparge plate, shown in Fig. 3-9, is less likely to plug than the sparge ring, and is much easier to clean in the event it does plug. Like the sparge ring, the sparge plate diameter should be 80% of the turbine diameter and no closer than two inches running clearance to the impeller blade bottom. Plate serrations promote uniform distribution of relatively small diameter bubbles of gas to enhance mass transfer.

The open pipe discharge of gas (Fig. 3-10) is ideally suited for slurries and/or liquids that are dirty. Of all the gas discharge assemblies the open pipe is least likely to plug up, and if it does plug can be unplugged by rodding out the pipe or unclogging with high pressure gas or liquid. The open pipe discharges large agglomerated gas bubbles that must be sheared into smaller bubbles and distributed uniformly by the gas–liquid mixing impeller. As such, mass transfer is relatively poor in comparison with sparge plates and rings.

Gas inlet pipe design should be based on minimizing gas side pressure drop to ensure proper presentation of volumetric flow rate of gas.

Each type of gas inlet assembly increases the fluid forces on the impeller due to the gas impinging on the impeller blades. This must be taken into account in the mechanical design of the agitator impeller, shaft and drive train. Nonuniform gas loading on the impeller can occur if the impeller and gas inlet device are not installed in symmetry or if the inlet device is not designed and located properly or if the gas rate varies frequently creating pulsing on the turbine. Not only can these situations increase mechanical load on the impeller and mixer, but process results can be adversely affected by poor gas distribution.

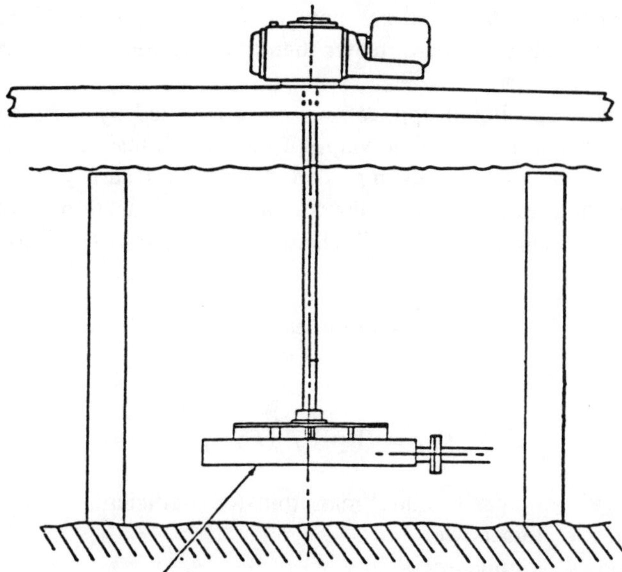

FIGURE 3-13. Submerged turbine aerator. (*Mixing Equipment, Inc.*)

Other Impellers

Radial flow impellers and axial flow hydrofoils with wide blades provide the best flow to head ratios for gas dispersion. Open impellers such as flat bladed paddle type impellers and narrow bladed hydrofoils should not be considered for gas dispersion. The solidity ratios are too low with these impellers. There is too much open space between blades and gas bypasses the high shear zones for gas dispersion. To compensate for this tendency for gas to bypass the blades of these open impellers, much more horsepower is required to disperse the gas. In addition the flow patterns of these types of turbines are not conducive to good dispersion. The natural tendency of the gas bubbles to rise in the liquid competes with the open impeller flow pattern.

For other gas-liquid applications, such as submerged turbine aerators for waste water treatment, other radial flow impellers and special spargers are used. An example of this is shown in Fig. 3-13. In this installation the impeller actually operates within the sparge ring. The sparge ring holes are distributed along the inside diameter of the ring, such that gas discharges inward toward the turbine blades.

E. MASS TRANSFER

Dispersion requirements for mass transfer of gas into a liquid or liquid–solid slurry are either uniform dispersion or at even greater power level (i.e., more than three times the isothermal gas power). For example, in fermenter applications power levels are much greater than that required for uniform physical dispersion.

In fact, most gas–liquid applications are controlled by mass transfer, not physical dispersion. The absolute value of the overall mass transfer coefficient (K_{Ga}) cannot be calculated or even predicted with any accuracy without running a pilot plant program. Mass transfer coefficient is a function of the physical properties of the materials involved. The overall mass transfer relationship may be expressed as

$$K_{Ga} = \text{lb moles/hr/ft}^3\text{/atm}$$

and

$$K_{Ga} \propto (\text{Hp}/V)^4 \, (F)^4 \, (\mu)^2 \tag{11}$$

where:
K_{Ga} = the gas to liquid mass transfer coefficient
Hp = horsepower
V = volume
F = superficial gas velocity = Q/A, ft/min
Q = volumetric gas flow rate, ft/minute
A = cross sectional area of tank, ft
x, y, z = exponents and are functions of the process variables
μ = viscosity of liquid

Mass transfer is often thought of in terms of the following relationship:

$$\text{Mass Transfer Rate} = K_{LA}(\Delta c) \tag{12}$$

where: K_{La} = mass transfer coefficient, hr^{-1}
Δ_c = concentration gradient.

It is not always possible to calculate the concentration driving force, Δc, in given liquids, since solubility data and reaction rate data are often needed.

A pilot plant program run with a variable speed mixer at various gas rates, with instrumentation to measure dissolved gas concentration in the liquid and gas concentration in the off gas, affords one the opportunity to identify the exponents x and y and K_{La} for scaleup criteria, etc.

Most liquid–gas mixing applications involve mass transfer. A frequent requirement is the dispersion of a certain quantity of gas per unit time. However, this is not usually the only mixing requirement. A more basic one is for a required rate of mass transfer, which should normally be the design basis for the mixer.

Basic measurement in terms of mass transfer is the mass transfer rate, usually expressed as the product of the volumetric coefficient and the average driving force of concentration.

There are numerous complexities in determining the proper mass-transfer driving force. Henry's law relates the partial pressure in the gas phase with the concentration in the liquid phase. The problem, however, is that Henry's law constants are not known for many gas–liquid mixtures. In such cases a decision must be made about whether to use a gas concentration or liquid concentration driving force.

It follows that the liquid mass transfer coefficient K_{La} is more suitable for liquid film controlled processes, while the gas mass transfer coefficient, K_{Ga}, is more suitable for gas film controlled processes. In concept this may be valid, but practically, with a Henry's law relationship, the only difference between K_{La} and K_{G_a} is the choice of concentration units.

In the periphery of the mixing impeller, the complex mechanism of mass transfer is not known. In this example, there is no reaction, or other mass transfer activity, acting to remove the dissolved gas from the process. If a large part of the mass transfer occurs in the impeller zone, the dissolved gas concentration may approach saturation, and any further increase in power may not improve process results.

Two driving forces can be defined as follows: (1) If the gas phase and the liquid phase are both well mixed, the concentration driving force will be the difference between the partial pressure of the exiting gas and the concentration of dissolved gas in the exiting liquid. (2) If the gas is unmixed and flowing through the liquid in plug flow, there will be a concentration driving force at the bottom of the tank and another at the top. The log-mean value of the two driving forces should be reasonably valid and accurate. There is no way to prove that one is better than the other without frequent sampling of the gas phase at various depths of the tank. Experience has indicated that the log-mean concentration driving force to be more consistent and practical for full size tanks.

F. EFFECT OF GAS RATE ON MASS TRANSFER

As gas rate increases at a given power level, pumping capacity becomes more important than fluid shear rate to optimize mass transfer. At low power levels, pumping capacity becomes still more important, while at high power levels, fluid shear rate in still the predominant factor affecting mass transfer. These findings are summarized in Fig. 3-14. Here we show the optimum D/T ratio (shaded area) for various ratios of gas flow to mixer power.

At high gas rates and lower mixer power levels, typical of a gas controlled regime, pumping capacity is more important than fluid shear rate for optimizing

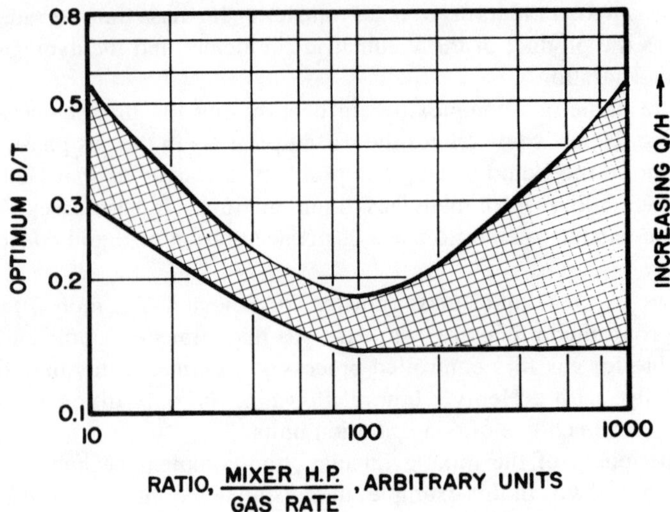

FIGURE 3-14. Optimum D/T ratios at various power to gas rates. (*Mixing Equipment, Inc.*)

mass transfer. This is reflected in the high *D/T* ratios. At higher power levels, typical of the mixer controlled regime, fluid shear becomes much more important, and the optimum *D/T* approaches relatively low values of 0.1 to 0.2. At extremely high power levels, the energy from the mixer is so great that the way it is distributed makes no difference, and there is no optimum *D/T* ratio.

Typical optimum *D/T* ratios in fermentation further illustrate that not everything is a mixing application can be optimized. The high shear rates produced by the optimum *D/T* ratios of 0.15 to 0.2 would be detrimental to biological organisms. Although *D/T* ratios from 0.35 to 0.50 are necessary for blending and heat transfer requirements in fermenters, as well as to prevent shear damage to the organisms, this range is not ideal or most effective for gas-liquid mass transfer.

Impeller Location

At a given gas rate and constant *D/T*, various arrangements of impellers, positioning of impellers and liquid levels show very little effect on gas–liquid mass transfer coefficient. This indifference is observed if the upper impeller (or only turbine in a single impeller installation) has at least 0.5*D* coverage of ungassed liquid level above it. At coverage less than 0.5*D*, the upper impeller will begin to splash surface liquid, simulating a surface aerator, and mass transfer will in fact be enhanced. A surface splashing impeller will not transfer oxygen as well as a submerged turbine with air introduced beneath it. In addition, it will pump very little.

In an unbaffled or partially baffled tank, vortex formation on the surface of the mixing contents will incorporate gas from the vapor space above the liquid back down into the liquid. Reentrainment of gas in this manner will also enhance the mass transfer coefficient. However, the negative aspects of a vortexing inpeller (and surface splashing impeller) are: (1) greater hydraulic loads imparted on the mixer shaft and drive train—the mechanical design of the agitator must compensate accordingly; (2) poor pumping and terrible flow pattern for top to bottom circulation.

Even though there are some minor differences in the flow to fluid shear rate ratios of multiple impellers versus a single turbine, the overall effect of impeller arrangement, position, and liquid level on mass transfer is minimal. This illustrates that, even though the instantaneous mass transfer coefficient around the impeller is much higher than it is throughout the batch, moving to different spots in the tank does not significantly change the integrated average mass transfer throughout the entire volume.

Therefore, even though Fig. 3-14 shows that D/T is very critical, the choice of multiple or single impellers, spacing, etc. is not critical. Impellers may be located to achieve other mixing results, such as solid suspension or blending, and still have effective gas–liquid mass transfer.

Bubble Diameter

A gas bubble of a certain diameter will rise through a liquid (or slurry) of a given viscosity at a particular rise velocity.

Table 3-1 lists bubble diameters for terminal rise velocities ranging from 0.3 to 0.0005 meters/second. As can be seen, as viscosity increases, bubbles must be considerably larger to flow upward at a given rise velocity. Impeller blade width should be a minimum of two to three times the size of the gas bubble being dispersed. Therefore, larger impellers are required for gas dispersion in higher viscosity liquids.

TABLE 3-1. Bubble Diameter Required to Give Various Terminal Rise Velocities.

Viscosity,	Bubble Diameter, mm, at Terminal Rise Velocity of:			
Pascal seconds	0.3 m/s	0.05 m/s	0.005 m/s	0.0005 m/s
0.001	1.0	0.5	0.13	0.05
0.01	2.3	1.0	0.25	0.10
0.1	5	2.5	0.75	0.25

Viscosity

All other things being equal, higher viscosity reduces the mass transfer coefficient.

Tank Shape

For a given volume of sparged air, the power required by both the air and the mixer is less with tall, thin tanks than with short, squat tanks of the same volume. Fig. 3-15 depicts this relationship. The greater pressure and higher superficial gas velocity in deep tanks provide process improvements that more than offset the extra horsepower required to introduce the air.

In general, axial flow impellers are more effective for solid–liquid suspension whereas radial flow disk turbines are more effective for mass transfer. It is not uncommon to install both radial and axial flow turbines on the same mixer shaft to optimize the balance of flow to shear ratio. The axial flow pattern is completely disrupted when gas is sparged into the tank. Only wide-bladed, high solidity hydrofoils have shown a propensity to disperse gas, while at the same time providing enough flow in the tank for flow controlled applications, such as solids suspension or heat transfer. When an *open* axial flow turbine (a wide-bladed hydrofoil, like a flat-bladed disk radial turbine, are considered *closed* impellers) is subjected to gas for dispersion purposes, the *open* turbine is rendered ineffective for either suspension, heat transfer, or mass transfer. The gas chokes the open impeller and prevents it from forming adequate flow velocity for any of the traditional flow controlled applications. For a combination gas–liquid–solid system, the radial flow and wide-bladed hydrofoil provide the proper flow to head (or shear) balance for process results.

Number of Impellers

Gas dispersed by one impeller at the tank bottom tends to coalesce rapidly into larger bubbles. Multiple turbines will maintain small bubble diameter, partic-

TANK HEIGHT, IN.	F, m/s	AIR POWER, kW	MIXER POWER, kW	TOTAL POWER, kW
3	0.0015	0.4	1.6	2.0
6	0.0030	0.75	0.75	1.5
12	0.0045	1.0	0.3	1.3

FIGURE 3-15. Power requirement for gas dispersion with various tank shapes. (*Mixing Equipment, Inc.*)

ularly for a relatively tall tank. Spacing can be farther apart than for flow controlled mixing applications, typically 1–2D apart, depending on tank geometry and number of impellers. Multiple impellers provide multiple high shear zones in the mixing tank to prevent coalescence of bubbles.

One method of dispersion with multiple impellers is to design the mixer such that most of the power is dedicated to the lower turbine for gas dispersion, while using less power in the impeller(s) above the lowest one for pumping, circulation, blending and solid suspension and heat transfer.

Experimenting with various impeller combinations and configurations will achieve the proper balance of flow, shear and power distribution.

A wide-bladed hydrofoil of high solidity will not have its flow pattern destroyed by the rising gas bubbles and hence gas dispersion and flow controlled applications can be achieved simultaneously. These mixing results can be achieved in open tank and draft tube circulation arrangements. Gas dispersion in a draft tube circulator is used for draft tube aeration of waste waters, etc. Here, since the flow from the turbine is confined within the tube, enough power is added by the mixer to overcome the upflow drive of the gas. A net downflow of gas bubbles to the bottom of the tank results, with effective mass transfer in most cases.

With a radial flow turbine that has no disk (i.e., blades emanating from a hub on the shaft) and is of ''open'' geometry, the gas can enter the low shear zone around the hub and pass through the impeller without being dispersed.

G. FERMENTATION

Agitators have been used in fermenters for many years. The functions of the mixer in fermentation processes are:

1. Provide uniform dispersion of gas bubbles
2. Produce small gas bubbles by shearing the inlet gas with fluid velocity gradients
3. Maximize retention time of the gas in the broth (liquid–solid slurry) by driving the gas bubbles to the bottom of the tank
4. Produce good bulk velocity and top-to-bottom turnover to create the driving force for function 3, enhance heat transfer, reincorporate nutrients from the surface froth back into the tank contents, and provide uniform nutrient and broth concentrations and temperature profile throughout the tank

The biological products of fermentations are living organisms with cell walls of limited resistance to fluid shear stress damage. The sensitivity of microbial cells to shear varies greatly. There has been great concern over the effect that fluid shear has on yield and productivity of various fermented products. When in doubt as to the threshold of shear sensitivity of a particular microbial organism, the fermenter mixer should be designed to minimize fluid shear.

Animal cell fermentations are encountered in biotechnical processes. The fermenter is often called a bioreactor. These cells do not have a cell wall per se, but are suspended in the fermenter medium as an encapsulation of fluid. Animal cells often must attach themselves to a surface for growth and reproduction. Inert solids have been successfully used to increase the surface area on which these cells can attach and grow. Depending on the size of the cell and surface irregularities in the inert carrier (microcarrier), the animal cells may grow either internally or exterior to the surface of the inert solid.

Excessive impeller shear rate can have two different, yet detrimental effects. The bond may be broken between the mammalian cells and the microcarrier or microcapsule, exposing the cells directly to high shear velocity gradients and potential damage in the fermented medium. Additionally, cell metabolic growth may be inhibited, reducing the ability to reproduce and thereby limiting the yield of production of fermented product.

Depending upon the scale being used in bioprocess reactors/fermenters, mixing systems other than conventional mechanical stirred agitators can be employed for shear sensitive applications. These include airlift agitators, fluid jet mixers, two-phase diffuser systems, etc. Care must be exercised in using these mixing systems to avoid cell damage caused by localized high shear zones at the gas (air) and/or fluid jet inlet points. Airlift mixers and fluid jet mixers are not as efficient as mechanically stirred agitators. Therefore, greater power must be introduced into the bioreactor to compensate for less efficiency, and this increased power could translate into detrimental shear effects on the cells. Scaleup criteria for airlifts, diffusers, and fluid jet mixers are not well established, whereas scale up is well known for mechanically stirred mixers. Extensive pilot plant testing and data gathering are required with airlift fluid jet and diffuser mixing systems to identify and quantify critical scaleup parameters.

Hydrofoil impellers in mechanically stirred bioreactors have been successfully applied to numerous shear sensitive mammalian cell fermentation applications. Once again, caution must be taken if hydrofoils are successfully used on a small scale basis to obtain critical scaleup data. In scaling up, care must be exercised to obtain the required blend time (bulk velocity) and shear rate, based on that achieved at the smaller scale.

Anaerobic and aerobic processes are involved in fermentation. Anaerobic processes require low oxygen uptake rates (low mass transfer rate of gas to liquid). The mixer design is controlled by mass transfer across the liquid–solid slurry boundary, possible effects of fluid shear on the organisms, and overall blending or bulk mixing of nutrients in the tank. Mammalian cell fermentation is an example of an anaerobic process.

Aerobic fermentation processes, on the other hand, require moderate to high oxygen transfer rates between gas and liquid phases. The controlling factor in mixer design is mass transfer across the liquid–gas boundary, usually accommodated by uniform physical dispersion of gas in the fermented broth or medium. Products of aerobic fermentation are not usually shear sensitive and heat transfer can often be a major consideration in mixer design. Unlike mammalian cell fermenters, which are typically small (10 liters to perhaps as big as 1,000 gallons in capacity), aerobic fermenters are much larger, with commercial scales typically ranging from 5,000 up to 50,000 gallons. Bacterial fermentations of *E. coli* strains and yeasts and mycelial fermentations to produce antibiotics such as penicillin, etc. are examples of aerobic fermentation.

Process and mixing requirements for various fermentation processes are tabulated in Table 3-2.

TABLE 3-2. Comparisons of Fermentations and Mixing Requirements.

Fermentation Type	Oxygen Transfer Rate	Shear Sensitivity	Mixing Requirements	Impeller Types
Anaerobic				
Mammalian cell	Low	High (need low fluid shear)	• Mild agitation • Little or no gas dispersion • Little or no heat transfer • Liquid-Solid Mass Transfer Controlling • Low Viscosity • Small scale • Shear sensitive	Low shear propellers or hydrofoils running at slow speed
Aerobic				
E. Coli bacteria Yeast	Moderate to high	Low (need high shear for gas dispersion)	• Vigorous agitation • Uniform gas dispersion • Heat transfer required • Gas–liquid mass transfer controlling • Low viscosity • Large scale • Not shear sensitive	Radial flow turbines or wide-blade hydrofoils or combinations thereof, at moderate to high speed
Mycelial Antibiotics	Moderate to high	Low (need high shear for gas dispersion)	Same as above for aerobic fermentations, but viscosity can be low to moderate	Same as above for aerobic fermentation

Mass Transfer in Fermenters

Sufficient oxygen mass transfer in aerobic fermenters is not usually a problem in a small scale pilot demonstration. Regardless of scale, comparable rates of mass transfer per unit volume are achievable on small and large capacity fermenters, assuming the proper mixer scaling criteria are identified and applied in scaleup or scaledown procedures.

A problem often encountered in scaling mixer performance relative to mass transfer rate comes from the blending phenomena. A large tank has a much longer blend time associated with it than the smaller tank, and oxygen deficiencies often result in remote portions of the tank contents. Mass transfer zones should be made a part of the overall mixing flow pattern in the entire system.

A small mixing tank has a much lower maximum shear rate in the impeller zone relative to large tanks. It might be required to use a nongeometric impeller and mixing tank on a small scale to duplicate maximum shear rate in the impeller zone and performance of proper variables in the larger scale.

In terms of minimum fluid shear rates required for some mammalian cell fermentations and other types of shear sensitive cells, a different approach to mixing requirement will be needed—i.e., it will be necessary to determine a threshold of acceptable shear rates with a particular impeller type in a demonstration scale, and duplicate that shear rate in the required scale, with perhaps a geometrically dissimilar impeller.

Flat bladed radial flow turbines with discs have been the workhorse in fermenters over the years. As a rule of thumb, 10 Hp per 1,000 gallon of ungassed broth with $D/T = 0.33$ flat-bladed radial flow turbines with disks achieved good results in large scale (10,000–50,000 gallon) aerobic fermenters for bacterial (yeast) and mycelial (antibiotics) fermentation of low viscosity. However, the need for bulk velocity and shear in proper balance for optimum process results, the introduction of shear-sensitive cell fermentations in biotechnology, and the increasing cost of energy in the late 1970s paved the way for the use of hydrofoils in fermenters.

It has been shown in pilot plant demonstrations and full scale fermenters that hydrofoils afford a better flow/shear balance, at reduced power requirements than the traditional workhorse flat-bladed radial turbines in fermenters. Various configurations of hydrofoils have been installed, primarily on a retrofit basis, whereby a hydrofoil(s) replaces a radial flow turbine(s). Usually in a retrofit campaign the mixer speed does not change, and the impeller diameter varies to either: (1) draw the same power and produce greater mass transfer or (2) produce the same mass transfer at reduced power draw (typically 40–50% less power required). It has been shown that wide-bladed hydrofoils with 4–6 blades of high solidity ratios are optimum design for dispersion of gas, shearing of gas bubbles and driving of the bubbles to the tank bottom—better than the flat-bladed radial flow turbine in most cases—and there is less shear, reducing damage to shear sensitive cultures.

Upper impellers in fermenters should be hydrofoils of narrow-bladed design for optimum results. These impellers need not be designed with gas dispersion in mind (the bottom impeller must be designed with gas dispersion in mind), but rather top-to-bottom turnover and bulk velocity to optimize blending, suspension and heat transfer. Some fermentations require a combination assembly of impellers—the lower one being a high shear radial flow turbine for greater shear while the upper ones can be hydrofoils of narrow blade design. Others are optimized with all impellers being hydrofoils (the bottom one a wide blade and the uppers of narrow blade design). Thus experimentation with various geometries of impellers, number of impellers, agitator speed, etc. should be conducted in fermenters to optimize mixing and process results and obtain initial scaleup parameters.

Oxygen is known to be a growth limiting nutrient for submerged aerobic fermentations. Oxygen is obtained by microorganisms through mechanical agitation and gas dispersion aeration of the fermentation medium or broth. Therefore it is very important to understand the mechanism of the transfer of oxygen for aerobic fermentation to optimize the design and scaleup of mixers in fermentation processes.

Aerobic fermentations are an intricate and complex number of different chemical constituents. Of importance is the need to understand oxygen transfer in the systems containing these individual components, before the highly complicated fermentation systems can be studied. The oxygen mass transfer process in fermentation is pictorially presented in Fig. 3-16.

The interest in hydrofoil impellers for gas–liquid processes and fermentation was precipitated from studies directed at retrofitting existing fermenter tanks and mixers. Typical power levels were 5–20 Hp per 1,000 gallons (10 Hp per 1,000 gallons was an average). Multiple radial flow disk impellers of $D/T = 0.33$ were used. Superficial gas velocities were typically 0.1–0.5 ft per second.

The hydrofoil impeller provides better mixing and process results (i.e., mass transfer/heat transfer). Radial flow impellers can create a staged mixing pattern with nonuniform oxygen and nutrient concentrations throughout the fermenter. In addition, damage to the microorganisms may occur due to high energy levels going into macro and microscale shear rates from the radial flow turbines. Replacing one or two of the radial flow impellers with hydrofoils, narrow bladed ones at the top of the vessel and wide bladed versions at the bottom, has proven to be optimum in many fermentation systems.

Hydrofoil impellers with wide blades and high solidity ratios (greater than 80%) have proven to be comparatively advantageous in most cases to radial flow impellers for gas dispersion and gas–liquid mass transfer.

Studies continue pertaining to mixer performance and process results with wide-bladed hydrofoils in fermenters. Several points can be made:

(1) The slope of the mass transfer coefficient as a function of power and gas rate has a marked break point and "knee" in the curve for a radial flow im-

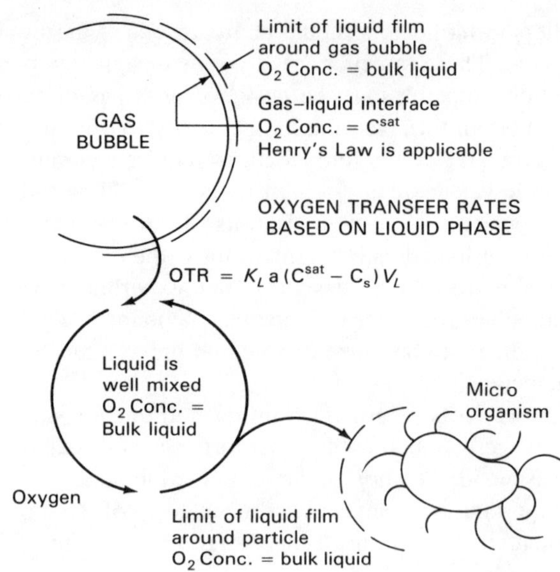

FIGURE 3-16. Oxygen gas–liquid mass transfer process. (*Prochem Mixing Equipment, Inc.*)

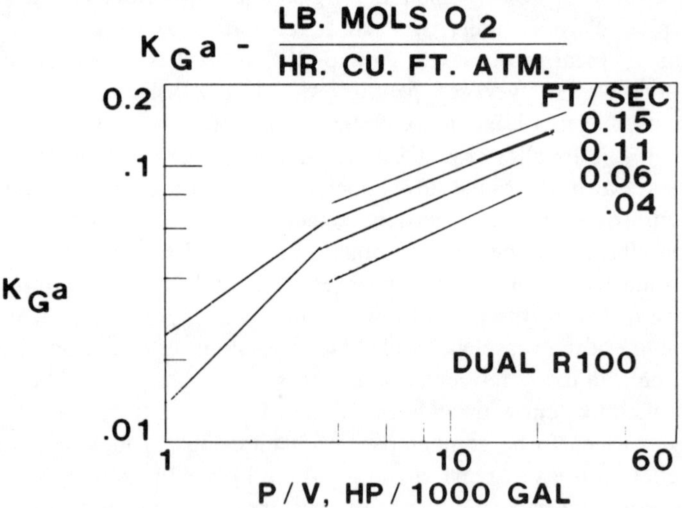

FIGURE 3-17. Typical plot of mass transfer coefficient K_{Ga} versus power per unit volume for various impellers and gas rates. (*Mixing Equipment, Inc.*)

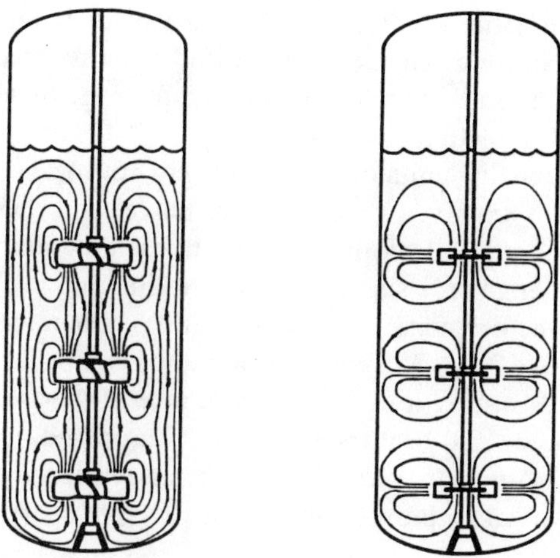

FIGURE 3-18. Flow patterns of hydrofoils and radial flow impellers in fermenters. (*Prochem Mixing Equipment, Inc.*)

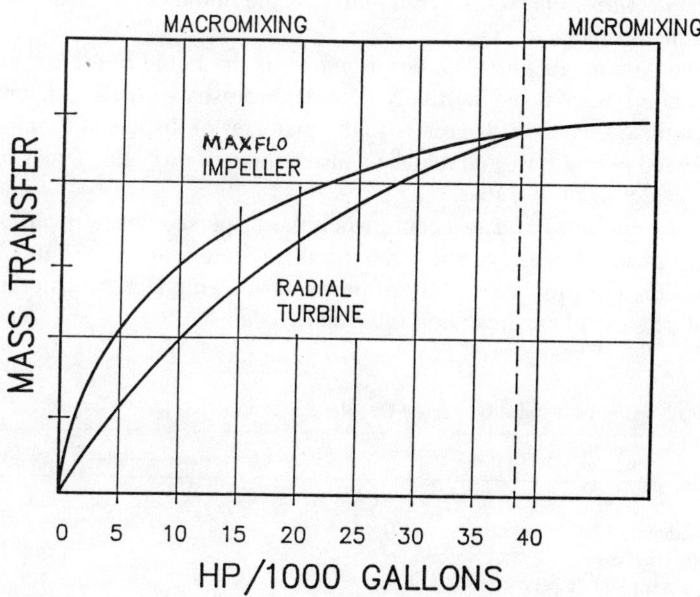

FIGURE 3-19. Relative improvement in mass transfer with hydrofoils at various power levels. (*Prochem Mixing Equipment, Inc.*)

peller. This characteristic is not shown by wide-bladed, high solidity hydrofoils. The slope is constant. Hence the hydrofoil is more effective, particularly at higher power input and/or greater mass transfer requirements. The relative performance of both impellers is shown in Figure 3-17.

(2) Retrofitting existing agitators is practiced to optimize mixing and process results. Typically this involves keeping mixer power and speed constant while retrofitting the impeller. Hydrofoils replacing radial flow impellers will always be of bigger diameter in this situation. As a result, one maintains essentially constant torque on the drive assembly and shafting. When the ratio of D/T for a radial flow turbine is 0.33–0.4, the gas–liquid mass transfer characteristics of the wide-bladed hydrofoil will be better than those of the radial flow impeller. Mass transfer, blending and heat transfer will be improved. Improved flow pattern will also result with the hydrofoils in fermentation service. This is depicted in Figure 3-18. Typical improvements in mass transfer are shown in Figure 3-19.

H. LIQUID–LIQUID EMULSIONS

A definition of terms pertaining to emulsions is necessary:

I. An *emulsion* is a dispersion of droplets or globules of one liquid into another immiscible liquid.

II. The *dispersed phase* (*internal phase*) is the liquid component that is dispersed in the droplets.

III. The *continuous phase* (*external phase*) is the liquid holding the dispersion. The continuous phase normally controls viscosity of the emulsion.

IV. Emulsion *particle size* refers to the diameter of dispersed droplets, usually expressed in microns, and may be a distribution of droplets. Types of emulsions are categorized in Table 3-3.

V. An *inversion of phases* occurs when the dispersed phase reverses to the continuous phase. Excessive shear rate by the agitator can cause this phenomenon to occur. For proper recovery of product following mass transfer and separation of phases, phase inversion must be avoided.

TABLE 3-3. Emulsions and Dispersed Droplet Size.

Description	Droplet Size (Microns)
Coarse	5–25
Milk with butter fat	5–10
Homogenized milk	1–2
Oil in water (well dispersed)	0.5–1.5
Microemulsions	<0.05
Above 1 micron sized droplets are macroemulsions.	

VI. *Surfactants* (surface active agents) are additives to emulsions designed to cause changes in the intermolecular forces of the liquids and thereby enhance the formation of optimum droplet size and/or size distribution in the emulsion.

VII. *Stabilizers* are substances added to an emulsion to increase the stability of an emulsion, usually by increasing viscosity, and improve mass transfer.

VIII. *Stability* is the ability of an emulsion, dispersed to a given droplet size distribution, to remain at that distribution long enough to allow mass transfer to occur achieving desired process results.

The movement of spherical liquid droplets in another liquid may be calculated by Stokes' law:

$$v = \frac{d^2\,(\rho_i - \rho_E)g}{18n} \tag{13}$$

where: v = settling velocity of the droplet, ft/sec
d = diameter of the droplet, ft
ρ_i = density of internal or dispersed phase, lb/ft^3
ρ_E = density of extremal or continuous phase
g = gravitational constant = $32.2\ \text{ft}/\text{sec}^2$
μ = viscosity, lb/ft · sec

The agitator must provide sufficient shear rate (velocity gradients) to generate a proper droplet size distribution to form an emulsion. Another role of the mixer is to provide sufficient bulk fluid velocity to overcome the natural movement of liquid droplets attributed to Stokes' Law. If this occurs, an emulsion is formed and maintained. When droplet movement by Stokes' law is the controlling factor, and the mixer is not driving the phases into forming an emulsion, phase separation or settling occurs.

Emulsification requires looking at the performance of the mixer from both a macro and microscale scenario. In a macroscale sense, blending of product throughout the mixing vessel to promote correct concentration, avoid settling and/or optimize heat transfer is required. These satisfy the flow controlled/bulk velocity controlled applications. From the perspective of the microscale, the mixer must provide a shear rate adequate to establish and maintain a certain droplet size distribution.

Typical steps in producing an emulsion are:

1. Prepare the continuous phase
2. Prepare the dispersed phase
3. Combine the phases and mix
4. Shear to desired droplet size distribution
5. Heat or cool as required

Batch processed emulsions can be accommodated by low, medium, high, or combination high/low speed agitators. Typical impeller types for the various speeds of mixing are given below.

Low Speed

1. Anchor agitators
2. Counter rotational drives
3. Hybrid impellers (i.e., fixed pitch axial turbine with anchor)

Low speed mixers for emulsions often have an anchor impeller (with or without wiper blades, but more often than not with wipers to clean the inside tank wall surface for optimum heat transfer) to provide bulk movement of the tank contents at high viscosity (greater than 100,000 centipoise).

Very often anchor impellers are installed in combination with paddle type impeller(s) of radial or axial flow pattern to maximize turnover of the tank contents and bulk motion, particularly with high viscosity emulsions (greater than 100,000 centipoise). These hybrid impeller arrangements may be mounted on a common mixer shaft, with a single drive assembly, and run at the same speed or speeds, in the case of a variable speed drive. Another configuration of hybrid impeller installation is to mount the impellers on independent, concentric shafting with dual drive assemblies "piggy backed" on the top (or bottom) of the tank. These drives can be counterrotating to each other and of different speeds.

Medium Speed

4. Marine propellers at 200–350 rpm
5. Axial flow impellers at 10–100 rpm:
   (a) Flat-bladed turbines of constant pitch
   (b) Hydrofoils of variable pitch
6. Radial flow impellers at 10–100 rpm:
   (a) Flat-bladed or curved impellers with disk
   (b) Flat-bladed or curved paddles
   (c) Flat-bladed or curved shrouded impeller

Medium speed impellers of the type indicated above are used for low viscosity emulsions, as they provide a unique and rather balanced proportion of flow and shear required for optimum mass transfer process results. Shear is needed for proper droplet formation and mass transfer in liquid–liquid emulsions. The use of flat versus curves blades is again predicated on the proper balance of flow and shear for the application at hand. Curved blades produce less shear, but more flow than straight, flat blades at the same power draw.

Shrouded radial flow impellers with straight or curved blades can provide modest lift or head development capability (usually on the order of 18–30" of head development or lift), in addition to providing proper flow/shear characteristics for emulsions and extractions. These type impellers for extraction/emulsion services on a continuous flow design premise are designed like a pump. In other words, the shrouded turbine is designed with a specific number, type (flat or curved), and geometry (width and length) of blade, and specific diameter and operating speed to produce a required flow and lift or head development. The head developed will equal the head required in the system due to actual liquid elevation differences, line friction loss, expansion and contraction loss, etc. The two liquid phases are drawn from settlers into an upstream or downstream mix tank for emulsification and extraction followed by settling and phase separation.

High Speed

7. High speed marine propellers
8. High speed dispersers or homogenizers
9. Rotor-stator mixers (mixer emulsifiers)
10. Recirculation with in-line static mixers:
    (a) Centrifugal pumps
    (b) Rotor-stator types
    (c) Colloid mills

High speed mixing for emulsification and extraction is pursued with low viscosity liquids requiring microemulsion with droplet sizes less than 0.05 microns. These high power, high shear machines produce far greater shear than flow for these applications.

Combination High/Low Speed

11. Anchor with medium shear paddle
12. Anchor with high shear paddle
13. Counterrotational drives:
    (a) With coaxial high shear
    (b) With high shear recirculation

The combination high and low speed mixing capability is ideal for emulsions of variable viscosity during formulation and production. Further flexibility is achieved by utilizing drive assemblies with variable speed features.

The method and sequence of addition of the components for emulsification is of vital importance. A controlled addition into the highest shear zone in the periphery of the mixing impeller prevents premature phase separation, and ensures appropriate mass transfer process results.

The mixer should be operable in only one phase upon startup. This prevents possible motor overload if the liquid of greatest specific gravity and/or viscosity is "seen" by the impeller during startups or upsets. Also this capability to operate in each liquid individually as well as in combination will prevent phase inversion with inverting emulsions.

Production of emulsions is complex and drop size correlations and size distributions are often not used in the design of agitators. For process analysis and study, it is helpful to understand how drop size will change with other mixing parameters. A key factor in analyzing emulsions is to determine whether or not the emulsion can be maintained once it is produced in the mixer.

The range of fluid shear rates in a large tank is considerably greater than that in a small one. Therefore the emulsion characteristics can change for each scale. This variation in fluid shear rate with scale is critical in designing emulsion-polymerization mixing systems. In this application emulsion droplet size must be achieved within a narrow distribution band to yield acceptable process results. The mixing fluid shear rate, consistent with achieving this droplet size distribution, must be the same at all scales to optimize process results.

Many experimental variables have been correlated, but the most important of these is that the mean Sauter drop diameter, $d_{\delta\mu}$, is inversely proportional to the droplet area and directly proportional to $N^{-6/5}$ and $D^{-4/5}$. The general relationship that follows is that, at equal particle volume, the average drop size and particle area in the tank will be the same. Knowing this relationship, one can calculate the value of the mass-transfer coefficient. Such calculations are useful, and in fact mandatory, in performing proper scaleup procedures to achieve similar process results in different scales.

The distribution of droplet sizes gives some qualitative idea of how emulsions vary in their range of particle sizes. A difficulty in making a complete calculation of droplet sizes for liquid–liquid dispersions is that there is a wide variety of shear rates in the system. There is a dispersing zone shear rate around the periphery of the impeller, as well as coalescence zone shear rates throughout the rest of the tank. The average shear is somewhere in between.

In producing a stable emulsion, one must look at the maximum impeller-zone shear rate, since all particles eventually reach that zone. If the emulsion remains stable, the particles have remained at the same size as that produced in the impeller shear zone.

I. EXTRACTION

In mixing applications, extraction is defined as the separation of one or more components of a mixture by the mixed addition of a solvent liquid. At least one of the components must be immiscible with or only partly soluble in the extractive liquid. At least two phases must be formed during and following the extractive process. Extraction processes are usually broken down into the following:

- Liquid–liquid extraction, in which the mixture treated is a liquid and the two phases formed are both liquid
- Leaching, in which one or more components of a solid mixture are removed by liquid treatment
- Washing, which is similar to leaching except that the solids removed are usually present only on the solid surface rather than throughout the solid phase
- Precipitative extraction, in which a homogeneous liquid system of two or more components is caused to split into two phases by the addition of a third component

In all of these systems, mixing is used to improve extraction rates by increasing contact areas and mass transfer coefficients. High shear and high turnover are generally provided to disperse the phases in liquid–liquid extraction and in leaching with power levels similar to dispersion. Extractions favor using radial flow turbines at modest tip speeds of 600–1,000 feet/minute. Leaching would favor an axial flow turbine or hydrofoil, depending on slurry viscosity, for proper flow and shear. However, washing and precipitive extraction usually require only mild agitation, similar to blending.

Extraction can be carried out in a single stage, mixed vessel, or in a series of tanks depending on retention time requirements for mass transfer, etc. Very often following the high shear mixing (required to establish an emulsion for extraction) will be settling of the immiscible phases. Settling action can be carried out in the same tank as the agitation by shutting the mixer off. This situation works well for batch operations.

Continuous flow extraction can be carried out in a series of mixing tanks followed by settling chambers. The mixing tanks will form the emulsion, whereas the settling chambers or "settlers" will allow the mixture to settle out into distinct immiscible phases based on a certain retention time. The phases can be separated and recovered by further processing.

Continuous extraction can be performed in vertical columns with multiple mixed stages. Each stage is perforated to allow flow to progress in either a cocurrent or countercurrent mode. (Inlet streams flow with each other in cocurrent flow or against each other in countercurrent flow.) Each stage has a mixing impeller in it and the stages act like mini mixing tanks. Shear rate/flow balance is usually adjusted by varying the agitator speed, since it is impractical to replace impeller type or change diameter of the individual impellers in each stage. The continuous extraction column can accommodate fairly high flow rates through relatively small mixing areas. Retention time is usually much less in a column extractor than in a conventional mixer-settler and therefore phase separation must occur faster in the column. See Fig. 3-20 for an illustration of a continuous flow, countercurrent extraction column.

Extraction processing requirements are so varied depending on the operation to be conducted that it is impractical to attempt to tabulate specific data required. Usually it is best to try to classify it under one of the other operations such as solids suspension or dispersion.

Liquid–liquid emulsion and extraction mixers require radial flow impellers to favor a heavier shear than flow balance and reduce cell to cell diffusion and backmixing. As such radial flow flat-bladed paddles and impellers with blades off a disk are excellent designs for emulsions and extractions.

Emulsions require a certain threshold of shear rate to form the proper droplet size distribution. However, excessive shear rate can create too fine a liquid droplet distribution. Phase separation problems can occur as well. Separation may not occur within a given retention time in the settler chamber or phase inversion can occur with an "oversheared" emulsion. Phase inversion occurs when the phases separate but flip-flop such that the lighter phase is on the bottom and heavier phase on top, rather than vice versa as in normal separation. It is usually advisable to pilot plant test to determine the threshold of shear rate required to create an optimum emulsion and phase separation with a given impeller type.

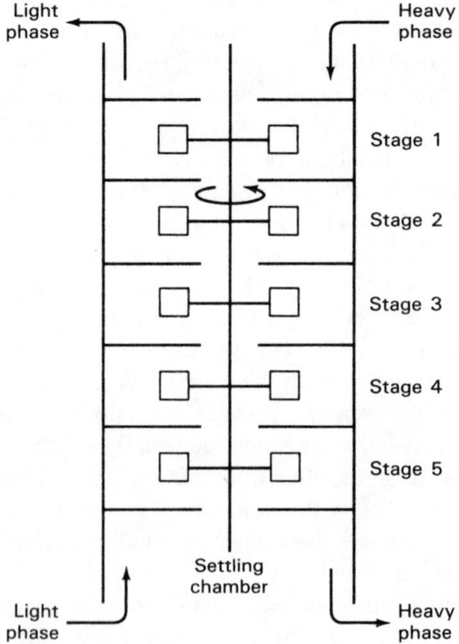

FIGURE 3-20. Continuous flow countercurrent extraction column.

Liquid–Liquid Mass Transfer

The rate of mass transfer between two immiscible liquids is given by the classic equation:

$$\text{Rate} = K_{La}\Delta c \tag{14}$$

where: K_{La} = overall mass transfer coefficient
Δc = concentration driving force.

In batch liquid–liquid processes, experiments are usually conducted by measuring the concentration of the solute in one phase and its transfer into another with the data typically plotted as K_{La} versus impeller speed. No general correlation has been found to provide coefficients, so experimental data are almost always required.

Scaleup to larger systems is usually based on constant power per unit volume, to maintain the same mass transfer coefficient. This requires that the emulsion have the same uniformity and dispersion on the two scales. However, maintaining an emulsion in a tank with a 10 foot liquid depth, for example, is much more difficult than doing it in a tank with a 6 inch depth. Care must be taken to determine the proper balance of head (shear) and flow with various impeller geometries in one scale to ensure desired process results upon scaleup to a larger scale.

J. DISPERSIONS OF SOLIDS IN LIQUIDS

Dispersions of solids in liquids describes the mixing of solids in liquids, into a pseudo-homogeneous mass that is more or less stable as measured by its life before noticeable separation occurs. These applications are often highlighted by both the requirements to suspend the solids in the liquid phase (flow controlled application) and prevent agglomeration of the solids and/or reduce particles to micron size (high shear application).

Solids dispersed in liquids can cover a wide range of product types from slurries to heavy dispersions such as pigment pastes, caulking compounds, etc. Power input per unit volume can vary widely. Conventional propellers or turbine impellers, at typical speed ranges for those impellers, satisfy the mixing requirements for some dispersions. In others, high speed impellers introducing higher shear and greater intensity of agitation are desirable to satisfy the dispersing requirement in a reasonable time frame. Some dispersions can be routine, while others may require running a pilot plant test program to obtain experimental data and scaleup criteria for optimum process results.

Additional data required to properly design the mixer for a dispersion of solids in liquid include:

- Type of dispersion (i.e., dispersion of coarse, medium, or fine grind solids in liquid).
- Relative amounts of each phase.
- Viscosity of liquid and final dispersed product, if known, together with details on temporary or interim viscosity conditions, if more extreme than initial or final conditions.
- Rate of addition of one component into another, and in what order.
- Some expression as to ease or difficulty of wetting of solids. Materials which are of a fluffy, light nature tend to float on the surface of the liquid. Others may tend to form agglomerates that resist complete wetting. Both conditions require a greater intensity of agitation to complete the dispersion.
- Solids particle size distribution.
- Specific gravity of both phases.
- Available time to create the dispersion. Where the solids content is low, solids are easily wettable, and agglomerates do not form, the application and horsepower requirements are similar to solids suspension. A change in available time has very little effect on horsepower levels, since the material is usually dispersed as rapidly as it is added. In more difficult dispersion mixing applications, horsepower levels and available time usually have a very definite relationship due to the need for a proper balance of high shear and adequate turnover (flow or bulk fluid velocity).
- Fineness of dispersion required to be produced by the mixer. Fine dispersions have micron sized particles. Some dispersions are considered complete when merely smooth in appearance (i.e., a paste). Others may require reduction of agglomerates to obtain maximum micron size. Agglomerates formed after initial entrainment of solids may be reduced rather easily up to a point, after which further reduction becomes exceedingly slow with traditional impellers at traditional power levels. Under this condition, if time is of the essence as far as making the dispersion, a special high shear/high power/high turnover mixer will be required. If subsequent processing for particle size reduction is planned for in other types of equipment (i.e., roller, sand or colloid mills), this should be stated since it will simplify the dispersion requirement of the mixer.

Impeller types and flow to shear balance are presented in Table 3-4 for various dispersions.

Solids that tend to float on the liquid surface (and are thereby considered difficult to "wet") can be incorporated into the high shear zone of the impellers for dispersing by creating a vortex on the surface. Solids are sucked into the impeller and dispersed in the liquid. Vortex formation can be had by running in a mixed tank without baffles or with partial baffles (i.e., baffles cut off in height at the top).

TABLE 3-4. Impeller Types for Various Dispersions.

Type of Dispersion	Impeller Type	Flow/Shear Balance
Slurries and coarse grind dispersions/low viscosity (i.e., solids suspension applications)	• Propellers • Axial flow turbines • Hydrofoils • Fixed pitch	Flow controlling
Medium grind size/low viscosity dispersion	• Radial flow turbines • Flat-bladed paddles	Balanced flow/shear
Fine grind size/low viscosity dispersion	• Bar turbine (disk with bar stock alternating top and bottom of disk)	High shear
Solids tend to agglomerate or particle size reduction required/low viscosity	• Sawtooth open disk impeller • Closed rotor/stator	Higher shear
Micron and submicron particle size dispersions	• Colloid mills • Homogenizers	Highest shear/low flow
High viscosity dispersions	• Anchors • Helical	Higher shear than flow Higher flow than shear

It is suggested that solids that are difficult to wet out, and/or tend to agglomerate and/or require size reduction, as well as dispersion, be mixed by an agitator with an axial flow turbine on top (creating a vortex for solids incorporation) and a high shear impeller (i.e., bar turbine, etc.) on the bottom. In this manner the proper balance of flow/shear in the mixed tank contents is achieved by dual turbines.

4

Mixer Scaleup

This chapter is devoted only to the scale-up of *mechanical* rotating agitators. Static mixer scaleup is not addressed here.

A. GEOMETRIC SIMILARITY

Scaleup is difficult at best when it comes to duplicating at a larger scale mixing and process results observed at a smaller scale. Some basic concepts will be presented in this chapter, and many cautions wil be tabled.

Experiments performed relative to scaleup determine the effects on mixing parameters as the size of the system changes. Fig. 4-1 shows that in order to maintain total similarity on scaleup, the larger tank would have to be made up of a series of smaller ones. In the larger volume tank in Fig. 4-1, higher bulk fluid velocities are required to provide the same circulation time as that seen in the smaller tanks.

Geometric similarity means that all pertinent dimensions are similar and have ratios constant for scale-up.

Kinematic similarity is maintained when all velocities have a common constant ratio.

Dynamic similarity occurs when all force ratios are constant. The four forces readily used for analysis and evaluation of scaleup criteria are: input force from the mixer, which is a function of impeller speed and diameter, and three opposing forces—viscosity, gravity, and surface tension. Dynamic similarity is predicated on the ratios of all four of these forces being equal for scaleup. However, the reality remains that for the same fluid, having the same density and viscosity (i.e., operating at similar temperature in both scales), only two of the four forces can be equal.

Geometric and dynamic similarity require the analysis and evaluation of di-

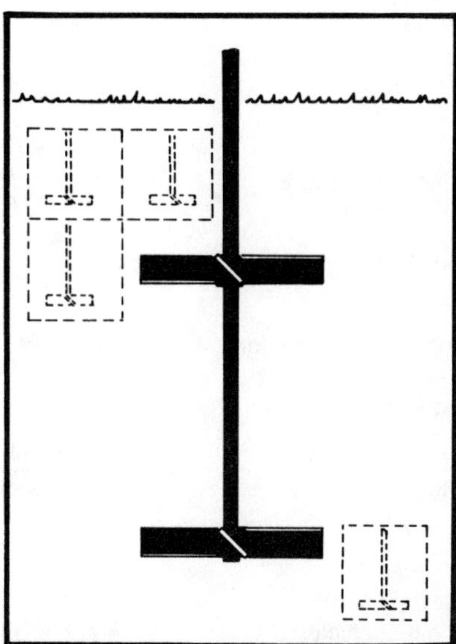

FIGURE 4-1. Large scale mixing vessel has different characteristics of shear rates, pumping capacities and circulation times compared to a small scale vessel. *(Mixing Equipment, Inc.)*

mensionless groups for scaleup. These dimensionless groups for various scaleup scenarios are summarized in the table in Fig. 4-2. The ratio of inertia to viscous force is expressed as the Reynolds number, the ratio of inertia to gravity force is the Froude number, and the ratio of inertial force to surface tension is the Weber number. These dimensionless numbers are useful considerations for scaleup.

For heat transfer scaleup, correlations are required. A process dimensionless ratio depicts the measured process result divided by the relative difficulty of carrying out the process. See Fig. 4-2. The heat transfer film coefficient divided by the thermal conductivity of the fluid, with a length term added to the correlation, is used.

In blending applications and scaleup scenarios, blend time relative to either the impeller speed or circulation time is used.

For fermentations, an adequate measure of process results that can be addressed as a dimensionless number does not exist. Numerous mixing applications cannot be expressed in terms of dimensionless numbers for scaleup, and another proven method must be used.

The following are represented in Fig. 4-2:

X is any pertinent dimension
F_i is inertia force

Hydraulic similitude

Geometric $\frac{X_M}{X_P} = X_R$

Dynamic $\frac{(F_i)_M}{(F_i)_P} = \frac{(F_V)_M}{(F_V)_P} = \frac{(F_G)_M}{(F_G)_P} = \frac{(F_\sigma)_M}{(F_\sigma)_P} = F_R$

Force ratios

$$\frac{F_i}{F_V} = N_{Re} = \frac{ND^2p}{\mu}$$

$$\frac{F_i}{F_G} = N_{Fr} = \frac{N^2D}{g}$$

$$\frac{F_i}{F_\sigma} = N_{We} \frac{N^2D^3p}{\sigma}$$

Application of hydraulic similarity to heat transfer

$$h = f(N, D, \rho, \mu, c_p, k, d)$$

$$\frac{\text{Result}}{\text{System conductivity}} = f\left|\frac{\text{Applied force}}{\text{Resisting force}}\right|$$

$$\frac{hD}{k} = \left|\frac{ND^2p}{\mu}\right|^x \left|\frac{C_p\mu}{k}\right|^y \left|\frac{D}{d}\right|^z$$

Where x, y and z are empirical coefficients

Application of hydraulic similarity to blending

$$\theta = f(N, D, \rho, \mu, T)$$

$$\frac{\text{Result}}{\text{System conductivity}} = f\left[\frac{\text{Applied force}}{\text{Resisting force}}\right]$$

$$\theta N \alpha \left[\frac{ND^2\rho}{\mu}\right]^x \left[\frac{D}{T}\right]^z$$

Where x, y and z are empirical coefficients

FIGURE 4-2. Parameters and dimensionless groups of use in various scaleups. *(Mixing Equipment, Inc.)*

F_v is viscosity force
F_G is gravity force
F_σ is surface tension force
N_{Re} is Reynolds number
N_{Fr} is Fronde number
N_{We} is Weber number
N is impeller speed
D is impeller diameter
ρ is fluid density
μ is fluid viscosity
g is acceleration due to gravity
σ is surface tension of fluid
h is mixer side heat transfer film coefficient
f means "is a function of"
c_p is fluid specific heat
K is fluid thermal conductivity
d is diameter of heat transfer coil or pipe
θ is blend time
T is tank diameter
$\propto$ means "is proportional to"

When one mixing variable is kept constant during scaleup, all others vary. Table 4-1 reflects what happens with certain mixing variables on scaleup, when

TABLE 4-1. Properties of a Fluid Mixer on Scale Up

Property	Pilot Scale 20 Gallons	Plant Scale 2500 Gallons			
P	1.0	125	3125	25	0.2
P/VOL	1.0	[1.0]	25	0.2	0.0016
N	1.0	0.34	[1.0]	0.2	0.04
D	1.0	5.0	5.0	5.0	5.0
Q	1.0	42.5	125	25	5.0
Q/VOL	1.0	0.34	[1.0]	0.2	0.04
ND	1.0	1.7	5.0	[1.0]	0.2
ND^2	1.0	8.5	25.0	5.0	[1.0]

one mixing parameter is kept constant in specific scaling situations. The parameters kept constant in separate scaleups are: power per unit volume, pumping capacity per unit volume, tip speed (π DN), and Reynolds number.

When power per unit volume is kept constant during scaleup, all other mixing variables change as noted in the same column in Table 4-1. In particular, pumping capacity per unit volume decreases markedly although overall pumping capacity increases dramatically. Scaling up mixer design based on constant pumping capacity per unit volume requires keeping impeller speed constant, as well as circulation time. In order to keep blend time constant, power per unit volume must increase proportional to the square of the tank diameter. Obviously, this is an impractical situation and therefore blend times in the case of large mixers are much longer than blend times of smaller equipment.

If the impeller tip speed (or maximum shear rate) is kept constant, the impeller or shaft speed must decrease, as does power per unit volume. Reynolds number will increase slightly, whereas pumping capacity per unit volume will drop. Tip speed is not usually a good scaleup parameter to keep constant and very often will create unsatisfactory mixing results, particularly those that are flow control (pumping) dependent. Holding Reynolds number constant during scaleup is not a suggested approach either. Power and practically all other scaleup parameters will decrease.

The best approach is to determine which parameters are important for mixing and then investigate and evaluate the various parameters in various scales of mixing to see how they might change (and therefore change mixing and process results) on scaleup. Changes that have a positive effect on process result should be identified and quantified. In a similar manner, changes that have a negative effect on process results should also be studied.

If scaleup is considered too risky or questionable, numerous small vessels may be incorporated in the process operations scheme, rather than fewer large volume tanks.

In a stirred tank, particles tend to catch up with each other only when they have different velocities. Shear rate is a measure of the velocity differential or gradient between particles, and is essential as a driving force for agitation. For a given velocity, more power is required when shear is present than when it is not.

Shear stress is the multiple of shear rate and fluid viscosity. For radial flow impellers, average shear rate around the periphery of the turbine increases with impeller speed. Average shear rate is unaffected as impeller diameter increases. Maximum shear rate (or that occurring within the impeller zone) is affected by speed and diameter (tip speed). Thus, maximum shear rate is proportional to the peripheral tip speed. On scaleup, operating shaft speed normally decreases and tip speed increases to achieve similar process results in both scales. For this reason, larger tanks have a greater spread of shear rates.

To understand what may be occurring in the process of scaling up, it is helpful to have some perspective as to "what is happening in a mixing vessel." Shear rates can be calculated by measuring velocity profiles at any point in the mixed contents, and by measuring the shear rate (velocity gradient) or slope of the curve.

A decision must be made as to whether one looks at the time average velocity at a given point, or if the instantaneous fluctuation at every point is going to be of interest.

Time average velocity determines values for shear rates between adjacent layers or "plates" of fluid acting largely on big particles, those on the order of greater than 200 micron in size. This gives us the classical definition of shear rate most often measured and quantified in a mixing tank.

Turning our attention to high frequency fluctuating shear rates in a turbulent flow regime, we see that these fluctuating velocities are caused by the passage of each blade of the impeller past our measuring point. These fluctuations are also due to the inherent turbulence in the stream. This can be continually characterized as the root mean square velocity fluctuation, which is relatively easy to measure yet difficult to interpret its importance.

The product of fluid shear and viscosity at any given point is the fluid shear stress. It is shear stress that actually implements and performs the work in mixing.

Pumping capacity is another important concept. No matter what shear rate is required for the process result, nothing will happen if the fluids are not circulated and "turned over" throughout the tank contents. The impeller creates primary flow, followed by the development of secondary flow, which is flow entrained by the impeller from its surroundings in the tank.

Superficial velocity is the total flow in the tank divided by the cross-sectional area. It does have relative meaning in terms of circulating time and other mixing variables, but it represents a very simplistic view of how a particle would behave in the system. Particles have circulation times that may deviate significantly from the average circulation time.

Scaleup is a discussion of the principles and fundamentals behind all types of mixing applications. Pilot plant testing investigates and quantifies specific data that provides the controlling or limiting factor for the overall process result.

A large scale mixing vessel exhibits a different scenario than several small scale mixing vessels, as shown in Fig. 4-1. The large stirred tank has different shear rates (greater maximum shear and lower average shear rate), greater pumping capacity, lower circulation times, as well as many other different mixing variables, than several small tanks adding up to same volume.

It is not desirable to assume a mixing variable for scale-up (as shown in Fig. 4-3) for each and every specific mixing process. Rather, a pilot plant program should be run to identify the controlling factors in scaleup, such that process results are in fact duplicated in more than one scale when the controlling mixing variable is held constant. Since there are only two independent impeller variables of the three (diameter, speed, and power), the choice of impeller type limits the process correlation to any two of these parameters. The third impeller variable is linked by the power curves, in which the power number versus Reynolds number relationship is a factor. Therefore, it makes no difference mathematically which two of the three impeller variables are chosen for process correlation.

Geometric and dynamic similarity can produce viable correlations for certain processes as shown in Fig. 4-3. This figure illustrates how process results may

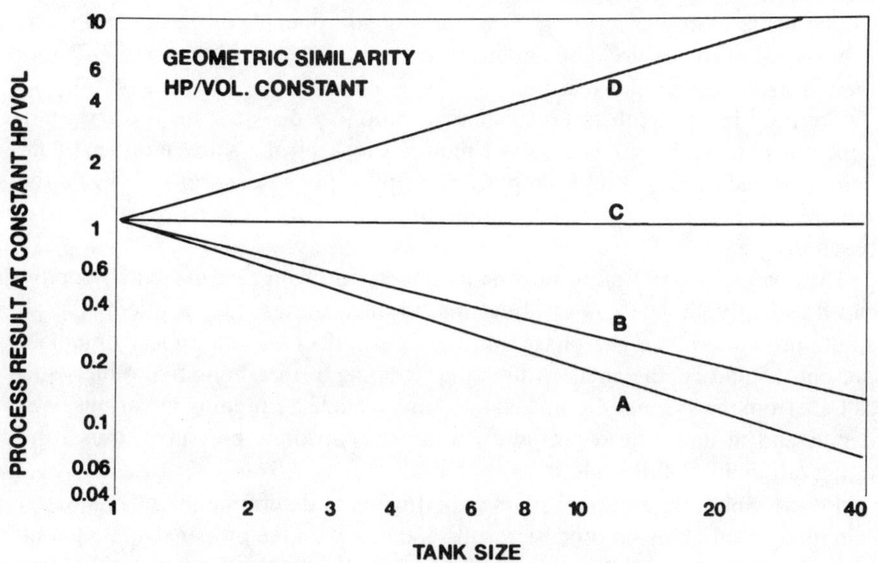

FIGURE 4-3. Tank size. *(Mixing Equipment, Inc.)*

change upon scaleup based on geometric similarity and constant power per unit volume. A and B represent decreasing process results, whereas C and D represent constant and increasing process results, respectively. Using geometric and dynamic similarity can produce very powerful relationships. However, it is extremely difficult to write a dimensionless group around a process relationship, and as such, the use of dimensionless number relationships is not particularly valuable or useful.

B. NONGEOMETRIC SIMILARITY

If a pilot plant testing campaign is to serve its purpose as a model of the full scale tank, the differences between small and large tank being mixed must be considered. The smaller ones have higher pumping capacities, shorter blend times, higher average shear rates, lower maximum shear rates, and smaller variations of shear rates. It is therefore impractical to have the pilot plant tank and mixer be geometrically similar to the full scale unit, since too many mixing variables will be different.

To compensate, one can resort to nongeometric similarity for modeling in the pilot plant. Impeller geometry can and should differ from that anticipated at full scale to bring as many mixing variables as close to being constant as possible. For example, the pilot plant impeller can have more or fewer blades than expected at full scale to bring pumping capacity, circulation, and blend time into line with that anticipated at the larger scale. At the same time blade width can be made narrower to bring shear rates down, once again in line with what is expected at full scale. The agitator speed can be made higher and D/T ratio made smaller for similar reasons.

Constant batch depth to tank diameter ratio Z/T does not need to be maintained at both scales. It is also not required to keep the same number of impellers at each scale. For example, if the pilot plant scenario has a Z/T that requires only a single turbine, multiple impellers may be pursued at full scale based on greater Z/T.

However, for gas-liquid dispersion service, a change in the batch depth Z can drastically change gas-to-liquid mass transfer rates. Depth affects the absolute pressure of the gas phase and may also affect any biological organisms present. Liquid depth can also affect the stripping of metabolic byproducts such as CO_2 from the system. Compensation can be made by running at various batch depths and/or under head pressure (to increase absolute pressure), thus simulating tall tanks at full scale.

Prior to initiating the pilot plant experiments to determine the effects of certain mixing variables on process results and ascertain the proper scaleup parameters, one must decide the types of experimental runs to be made (i.e., geometric versus nongeometric similarity), the approximate mixer speeds, and the data to be collected. Each process is different (i.e., the methods for evaluating experimental runs for a polymerization are quite different from those used in gas–liquid dispersions); therefore, it is impossible to list a simple series of experiments, common to all mixing applications, covering all possibilities.

Simple analysis of complex mixing problems can often be confusing. Understanding the complexity of the mixing system will help to formulate experiments in the pilot plant that are usable.

Nongeometric methods are needed to determine such variables as blend time, circulation time, and other indicators of what is actually happening in the tank. Nongeometric similarity in mixing experiments compensates for the way geometric similarity allows too many mixing parameters to vary. It is a way to truly simulate what goes on in the larger tank relative to mixing. Very often the pilot plant system will have an agitator that appears to be operating with poor physical mixing, to get meaningful studies and data and simulate a certain mixing variable(s) in the full scale. Unless the nongeometric similarity approach is taken, the effect of some mixing variables will be missed completely at the small scale. This would lead to mixing and process results in the large tank being considerably different from those observed in the small scale.

C. SCALEUP CRITERIA FOR BLENDING AND SOLIDS SUSPENSION

Impeller pumping capacity per unit volume Q/V, is a good indicator of the measured circulation time of a particle in a small tank of approximately 50 gallon capacity. For greater volumes of up to 1000 gallon capacity, circulation time is greater than Q/V. In general circulation time is proportional to blend time.

Larger tanks generally have a lower Q/V value than do smaller tanks and therefore the calculated and actual measured circulation times are longer. In addition, for a larger tank with a mixer at a given speed, the deviation of blend time around an average value is greater. Particles tend to trace a random flow path around a tank and circulation time corresponds to what would be expected for several smaller tanks in series, in lieu of one well mixed larger tank.

Initial consideration should be given to constant power per unit volume, P/V, and geometric similarity for scaleup of any mixing application. Power, speed, and diameter of the impeller (keeping D/T constant on scaleup adhering to geometric similarity), and the shear rates should all be estimated. A judgment should be made as to how these variables will affect mixing and process results. Other parameters that might be important pertaining to process results should be considered.

Scaleup on constant power per unit volume and geometric similarity will produce conservative scaleup criteria, even though blend time and circulation time will be longer at full scale than at demonstration pilot scale. If the process result is a function of the number of times that a particle or material passes through the high shear zone of the mixing impeller, the cycle time may be important.

Experimentation has shown that for solids suspension mixing applications, with free settling solids (i.e., solids with terminal settling velocity greater than 1 foot/minute), the exponent on power per unit volume for scaleup is less than 1.0. In fact, to achieve similar degrees of suspension in two different scales, a factor of $(P/V)^{0.95}$ can be used for free settling slurries.

In a similar manner, to duplicate mixing and process results at different scales with hindered settling slurries (i.e., solids with terminal settling velocity less than 1 foot per minute), a factor of $(P/V)^{8/9}$ can be used. In both cases of free and hindered settling scaleup criteria, geometric similarity has been assumed (i.e., equal D/T, and same impeller type).

To modify some of the effects of pumping capacity, blend time or shear rate, different D/T ratios at full scale should be looked at. The law of conservation of momentum states that fluid momentum is constant at all points in the mixing vessel (i.e., momentum = flow × velocity). Therefore as flow increases, with larger D/T impeller, velocity will decrease in order to keep the product of flow times velocity, or momentum constant. Also power P is proportional to flow Q times velocity head H, or $P \propto QH$, and as flow increases at constant power, velocity head must decrease commensurately.

Larger D/T impellers tend to have a smaller variety of shear rates than do smaller ones. Impellers located closer to tank walls have more uniform shear rates. Larger impellers running at slower speed to achieve desired process results will do so at less power requirement, but at greater torque for the power selected. This translates into a larger mixer drive train (speed reducer).

D. MASS TRANSFER SCALEUP CRITERIA

If the process is controlled by mass transfer or chemical reaction, physical velocity and uniformity criteria may prove to be an incorrect scaleup criteria.

As an example of mass transfer scaleup, consider aerobic fermentation. The general equation for mass transfer rate is:

$$R = K_{La} \Delta C \tag{1}$$

where: R = mass transfer rate, lb mole/ft^3 – sec
K_{La} = mass transfer rate coefficient, sec^{-1}
Δ_C = concentration driving force, lb mole/ft^3

The only variable that the mixer controls is K_{La}. A pilot plant demonstration program should be run with various types of impellers at various operating speeds, and with various gas rates or superficial velocities. Instrumentation should be incorporated to measure dissolved oxygen in the tank contents and gas analysis equipment to measure incoming or outlet gas composition. With this setup, a matrix of experiments should be conducted to evaluate mass transfer as a function of mixer horsepower and superficial gas velocity.

Mass transfer coefficient K_{La} can be calculated from Eq. (1) based on measured concentrations of oxygen in the liquid and off-gas and based on a mass balance calculation to determine mass transfer rate R.

Superficial gas velocity can be calculated based on measured gas rate and cross-sectional area of the tank. Horsepower can either be measured with a torque cell or dynamometer or calculated based on measured speed.

A plot of K_{La} versus superficial gas velocity F and K_{La} versus horsepower per unit volume should produce a straight line plot on log-log paper, the slope of which defines an exponent for scaleup.

Figures 4-4 and 4-5 show typical plots of mass transfer coefficient as a function of power and gas velocity.

Assuming a desirable K_{La} is achieved in the pilot plant, one can achieve the same mass transfer coefficient in the larger scale by using the following ratio analysis.

$$K_{La1} \propto (\text{Hp/Volume})_1^x \, (F)_1^y \tag{2}$$

$$K_{La2} \propto (\text{Hp/Volume})_2^x \, (F)_2^y \tag{3}$$

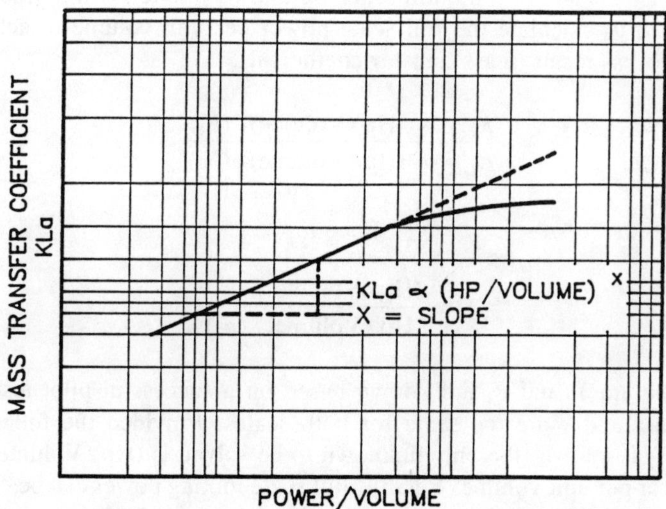

FIGURE 4-4. Mass transfer as a function of power. *(Prochem Mixing Equipment)*

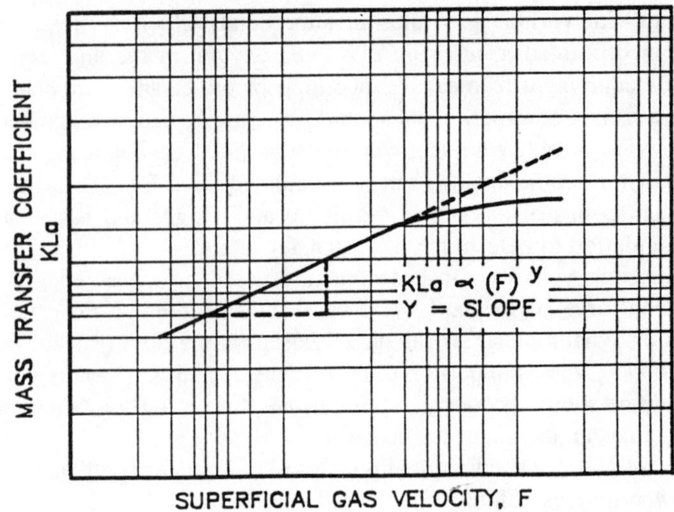

FIGURE 4-5. Mass transfer as a function of gas velocity. *(Prochem Mixing Equipment, Inc.)*

Subscript 1 is for the pilot plant, subscript 2 is for the larger scale. Exponents x and y are slopes of straight line plots of K_{La} versus Hp/Volume and K_{La} versus F, respectively, on log-log paper.

Since the desired full scale mass transfer coefficient K_{La} is the same as that achieved in a successful pilot program, and provided that Hp/Volume and F in the pilot scale and F in the full scale are known, the following ratio analysis can be used to calculate the full scale power per unit volume to achieve the desired process result (mass transfer coefficient).

$$\frac{K_{La1} \propto (\text{Hp/Volume})_1^x \, (F_1)^y}{K_{La2} \propto (\text{Hp/Volume})_2^x \, (F_2)^y} \tag{4}$$

but since $K_{La1} = K_{La2}$

$$1 \propto \frac{(\text{Hp/Volume})_1^x \, (F_1)^y}{(\text{Hp/Volume})_2^x \, (F_2)^y} \tag{5}$$

$(\text{Hp/Volume})_1$ and F_1 are known based on a successful pilot test and the exponents x and y are the same for both scales. Provided the full scale gas velocity F_2 is known, the only unknown to be solved is $(\text{Hp/Volume})_2$ or full scale power per unit volume. Finally, full scale mixing power can be calculated by multiplying $(\text{Hp/Volume})_2$ by full scale ungassed volume.

The exponents x and y should be determined based on pilot testing. Literature values can be used without testing to confirm the exponents, but the risk is great. Overestimated exponents can result in oversized (overpowered) agitator to achieve desired process results. Underestimated exponents can result in too small a mixer (underpowered) to achieve desired mass transfer. An overpowered agitator will not be cost effective from an operating and capital cost perspective. An underpowered agitator simply will not produce the desired mass transfer process results in the large scale. Productivity will be insufficient and the consequences of reduced yield could be disastrous.

The following is an example of a scaleup calculation for a mass transfer application to achieve equal mass transfer.

Assume that an 18,000 gallon fermenter drawing 10 Hp is dispersing air rising at a superficial velocity of 0.05 ft/second and a satisfactory mass transfer rate occurs.

The same mass transfer is desired in a 180,000 gallon fermenter. However, at equal air volume rate per liquid volume to that in the smaller vessel (for stoichiometric equivalence), the superficial rise velocity in the larger vessel is 0.12 ft/second. What is the mixer power required in the larger fermenter to achieve equal mass transfer?

From Eq. (5),

$$\frac{K_{La1}}{K_{La2}} = 1 \propto \frac{(\text{Hp}/\text{Volume})_1^x\,(F_1)^y}{(\text{Hp}/\text{Volume})_2^x\,(F_2)^y}$$

Assume that the exponents x and y have been determined from testing to be 0.5.

$$1 \propto \frac{(10/18{,}000)^{.5}\,(.05)^{.5}}{(\text{Hp}/\text{Volume})_2^{.5}\,(.12)^{.5}}$$

Rearranging and solving for $(\text{Hp}/\text{Volume})_2$:

$$(\text{Hp}/\text{Volume})_2 = .00023$$

therefore

$$\text{Hp}_2 = (.00023)(180{,}000) = 41.33\ \text{Hp}$$

For equal mass transfer, a mixer power draw of 41.33 Hp in the 18,000 gallon fermenter will suffice. However, one should check to see if the same dispersion is achieved (i.e., uniform or intimate dispersion will be required).

Note that if scaleup were done on an equal power per unit volume approach, the larger fermenter mixer power would have calculated to be 100 Hp. This would have lead to a much larger mixer then required for equal mass transfer.

E. HEAT TRANSFER SCALEUP

All the mixing variables and parameters that can be used to describe the flow pattern and mixer performance for scaleup criteria are numerically different from a small to a large scale mixing tank. It is imperative to carefully consider the effect of each and every variable on scale-up and to be certain that, if certain minimum and maximum values of any particular quantity are required, that proper adjustments be made to the other mixing variables.

In many heat transfer mixing applications, another mixing requirement, say suspension of solids, may control the mixer selection and scaleup criteria, for which established guidelines exist. Without any other guidelines to go by, the scaleup procedures of other flow controlled mixing applications (i.e., constant Hp/Volume for blending, $(\mathrm{Hp}/\mathrm{Volume})^{0.95}$ being constant for scaleup of free settling slurries, and/or $(\mathrm{Hp}/\mathrm{Volume})^{8/9}$ being constant for scaleup of hindered settling slurries) can be extrapolated to heat heat-transfer mixing operations.

For equal process results in blending applications, mixer horsepower can be scaled up directly by batch volume (i.e., equal horsepower per unit volume), assuming geometric similarity. Therefore, for heat transfer applications, constant power per unit volume can be used to achieve approximately the same heat transfer coefficient for the same type of impeller and constant D/T. This method provides merely an approximation on scaleup, because the effect of power on the heat transfer coefficient is relatively small (i.e., $h \propto \mathrm{Hp}^{0.22}$, where h is the mixer side heat transfer film coefficient and Hp is mixer horsepower) and even a moderate error in scaling up will have only a minimal effect on heat transfer.

Heat transfer surface area per unit volume is a more significant factor in scaling up than power. For example, in a jacketed tank the heat transfer area per unit volume usually decreases on scaleup. To maintain the proportionate heat transfer per unit batch size, additional heat transfer surface area (such as coils internal to the tank or external heat exchange surface) would be required. Alternatively, other variables, such as temperature driving force (Δ_T), may have to be adjusted commensurate with the smaller heat transfer area.

Preliminary estimates of mixer side heat transfer film coefficients may be made by using Fig. 4-6. It shows typical mixer side heat transfer coefficients for heating and cooling of organic fluids. Use of this graph will provide order of magnitude estimates of coefficients suitable for preliminary design and mixer selection. The coefficients will obviously change, depending on the specific characteristics of the system and the mixer selected. The values are based on vertical tubes or a single bank of helical coils. For fully baffled jacketed tanks, the coefficients should be taken as 65% of the value obtained from Fig. 4-6.

The coefficients of aqueous solutions of inorganics will be three to four times higher than those from Fig. 4-6 because of higher thermal conductivities of these solutions.

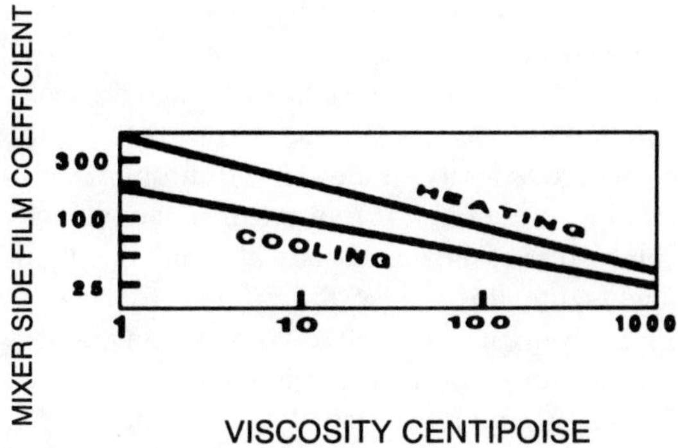

FIGURE 4-6. Typical mixer-side film coefficients for heating and cooling organic chemicals. (Thermal effectiveness 1.54.) *(Mixing Equipment, Inc.)*

As an option to estimating, mixer side heat transfer coefficients, h_i, can be *calculated* based on correlations presented in Chapter 2.

F. MACRO- AND MICROSCALE MIXING

A major concern pertaining to fluid mixing processes revolves around scaleup. The role of macro- and microscale agitation is better understood today and continues to evolve. Geometry, and possibly geometric similarity, and the roles they play in selecting pilot and commercial scale mixing systems, with similar mixing process characteristics, are also better understood today. Study in this area as well as work on numerical fluid mechanics, with an understanding of the role of energy spectrum, has led to successful modeling and scaleup of reactors.

All of the power transmitted by a mixer to the tank contents through the impeller appears as heat ultimately, by way of viscous dissipation. Power is converted to heat through viscous shear at the rate of 2,500 Btu per hour per horsepower. Viscous shear is only present in turbulent flow regimes and exists at the microscale level. Consequently power per unit volume is a major component of microscale mixing. The boundary limit for microscale mixing occurs when particle or droplet size is much less than 100 micron. At 1 micron particle size, in fact, it does not matter what specific impeller design is chosen to transmit the power, for the flow regime will be turbulent and microscale mixing will predominate.

Power per unit volume within the periphery of the impeller (which is typically about 5% of the total tank volume) is about 100 times higher than the power per unit volume in the balance of the vessel. The root mean square velocity gradient in the turbine zone is roughly 5–10 times greater than that in the rest of the tank. This has been confirmed by experimental work with sophisticated laser velocity measurement instrumentation.

For many open impellers (i.e., constant pitch and variable pitch hydrofoil axial impellers, propellers, etc.), the ratio of the root mean square velocity fluctuation to the average velocity in the mixing turbine zone is approximately 50%. However looking at the ratio of root mean square velocity gradient to the average velocity in the rest of the vessel indicates that this figure is on the order of 5–10%. Coincidentally, this is also the ratio of root mean square velocity fluctuation to the mean velocity in pipeline flow. Mixing tanks with mechanical agitation can produce microscale mixing phenomena not encountered in pipeline reactors. The process requirements will dictate whether this is good or bad.

Figure Fig. 4-7 depicts velocity versus time gradients for three different types of impellers—narrow bladed hydrofoils, constant pitch axial flow turbines and radial flow disk turbines. As can be seen, the differences between these impeller geometries are quite significant and can be important factors in reactor process results.

There are many fluid mechanics spectra applicable to process design and mixing results; however, it is impossible to specify and/or quantify these and apply them to mixer design. Industrial mixing processes are too complex to define required process results in terms of these fluid mechanics parameters.

Without question, the most practical aspect of these fluid mechanics studies has been the ability to design pilot plant experiments and programs (and in some cases full scale experiments) to establish process sensitivity to macroscale mixing variables (i.e., power, pumping, capacity, impeller diameter and speed,

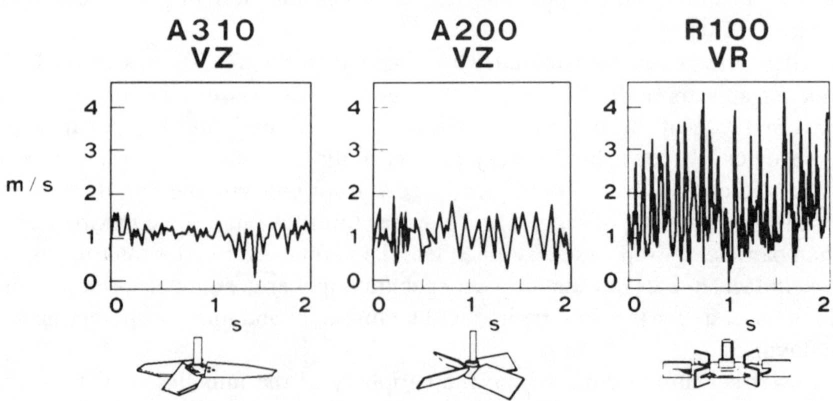

FIGURE 4-7. Comparison of velocity versus time for various impeller geometries at constant total impeller pumping capacity. *(Mixing Equipment, Inc.)*

macroshear shear rates, etc.) in comparison with microscale mixing parameters (i.e., power per unit volume, root mean square velocity gradients, and an estimation of microscale eddy size).

As previously stated, using geometric similarity, the macroscale variables can be summarized as follows:

1. Blend and circulation times will be much longer in the full scale than small scale (pilot) tank.
2. Maximum impeller zone shear rate will be greater in the full scale, while average shear rate throughout the tank contents will be less than the corresponding values in the small scale. Therefore, a much greater variation in shear rates exists in the full scale.
3. Reynolds numbers in the larger tank will be higher, typically running 5 to 25 times greater than those in the small scale. Thus, the flow regime is more likely to be turbulent in the full scale even though transitional or laminar in the small scale.
4. Large tank flow patterns tend to be recirculatory from the impeller through the tank back to the impeller. This simulates a number of individual tanks in series. The mean circulation time is increased over that predicted from the impeller pumping capacity. Likewise, the standard deviation of the circulation times around the mean also increases.
5. Sufficient heat transfer is usually more difficult to achieve in the full scale. In-tank surface area for heat transfer in the form of helical coils, vertical tubes, panel coils, etc. can create resistances to flow and dead spots of low or no circulation. This is why external heat transfer surface is often used.
6. In gas–liquid mass transfer applications, gas superficial velocity usually increases upon scaleup, if constant stoichiometric proportions of gas rate to liquid volume (i.e., equal VVM) are maintained. This increased resistance to flow will further increase circulation time in the larger scale.

Microscale phenomena depend largely on the energy dissipation per unit volume, although energy spectra are also of concern. As mentioned before, the energy dissipation per unit volume in the mixing zone "seen" by the impeller is approximately 100 times greater than that occurring in the rest of the tank. The result is a ratio of root mean square velocity gradient to average velocity on the order of 10:1 between impeller mixing periphery and the rest of the tank.

The approach to pilot plant testing and scaleup is predicted by the specific process and required process results. No set of rules can be laid out to guide one through the myriad possible test campaigns. A few guidelines that can be followed in carrying out a successful pilot plant program are:

1. For a given process look qualitatively at the possible role of fluid shear rate and shear stress. Evaluate the affect, if any, fluid shear rate and shear stress have on process results. If there is no effect, then these extremely complex fluid mechanics can be dismissed and the testing can focus on process design based on macroscale variables such as uniformity, circulation time, blend time, or velocity variables. Usually flow controlled mixing applications fall into the scenario (i.e., blending, suspension of solids, and heat transfer), unless shear rate/shear stress is critical in addition to flow controlling criteria.
2. Assuming fluid shear stresses are significant factors in obtaining a successful process result, one must qualitatively evaluate various scales (volumes) against which shear rate/shear stress influence results. If particles, bubbles, droplets, or fluid clumps are on the order of 1,000 microns or larger, the mixing variables will be macroscale and average velocities (not root mean square gradients) at any given point in the tank contents will dominate the mixer design and process results.

Every geometric design variable can affect the role of shear stresses when macroscale mixing variables are involved. Power, impeller speed and diameter, blade shape and width or height, thickness of material used to make the impeller, number of impeller blades, turbine location, baffle location and number of impellers are examples of macroscale mixing variables that should be investigated in the small scale to determine their effect on shear stress and process results.

Microscale mixing variables will dominate process results when particles, droplets, bubbles, or fluid clumps are on the order of 100 microns or less in size. In this case the critical parameters usually include power per unit volume (energy dissipation per unit volume), distribution of power per unit volume between the impeller and the rest of the tank contents, root mean square velocity fluctuations, energy spectra, eddy size in the wake of the impeller blade passage for a particular power level, and fluid viscosity.

In order to make the pilot plant simulation more like a commercial scale unit in macroscale characteristics, the pilot plant mixing impeller must be designed such that the blend time is lengthened and maximum shear rate is increased over that traditionally observed. This will yield a greater range of shear rates than normally found in the pilot plant, and will closely resemble the macroscale considerations to be obtained in the larger scale.

These conditions can be met using smaller D/T ratios and narrower blade widths than normally used in traditional pilot plant impellers. Care must be exercised when using the same impeller type in both the pilot and full scale, since it may not be possible to come close to the long blend time to be experienced in the commercial scale in the small scale. Going to a different impeller geometry in this case may be the answer. For example, radial flow impellers can be an excellent model in a pilot plant for axial flow turbines in the full scale.

Flow efficient hydrofoil impellers will produce a shorter blend time than other impeller types. The outstanding performance of hydrofoils may not stand out in pilot plant testwork, primarily because a pilot unit usually is already too good a blending mixer compared to what really happens in the larger scale.

A pilot plant should be at least of minimum size to accurately measure data and quantify process results. From the viewpoint of minimum blade width based on particle size, consider the following. Assume for example that the blade width is 1 centimeter. If the maximum impeller zone shear rate at the boundary of the discharge stream has a value of 10 reciprocal seconds, then the shear rate across ⅛ cm would be 9.5, across ¼ cm it is 7.0 and across ⅝ cm it would be 5.0 reciprocal seconds. A shear rate of 5.0 is also roughly the average impeller zone shear rate.

The shear rate across a 1 cm particle would be zero, since the fluid velocity on both sides of the impeller blade is the same and shear rate is defined as $\Delta V/\Delta r$ or velocity difference divided by incremental blade width difference. This also indicates that a small particle would see a shear rate of 10 (maximum impeller zone shear rate), while a big particle of say 1 cm size would see a shear rate of zero. A general and very practical rule of thumb is that the impeller blade height or width of the impeller discharge stream should be at least four times bigger than the largest particle to be mixed in a process.

Remember that no mixing application is purely macroscale or microscale mixing controlled. By the same token, no impeller design or configuration will produce purely macroscale or microscale mixing.

Example of macroscale mixing controlled processes are:

1. Blending and storage
2. Solid suspension
3. Heat transfer
4. Crystallization
5. Fermentation

Examples of microscale mixing controlled processes are:

1. Mass transfer
2. Liquid–liquid extractions
3. Dispersions
4. Gas dispersions (exclusive of fermentation)

Macroscale mixing of the bulk contents of a tank can be broken down into components of microscale agitation. This is illustrated in Fig. 4-8.

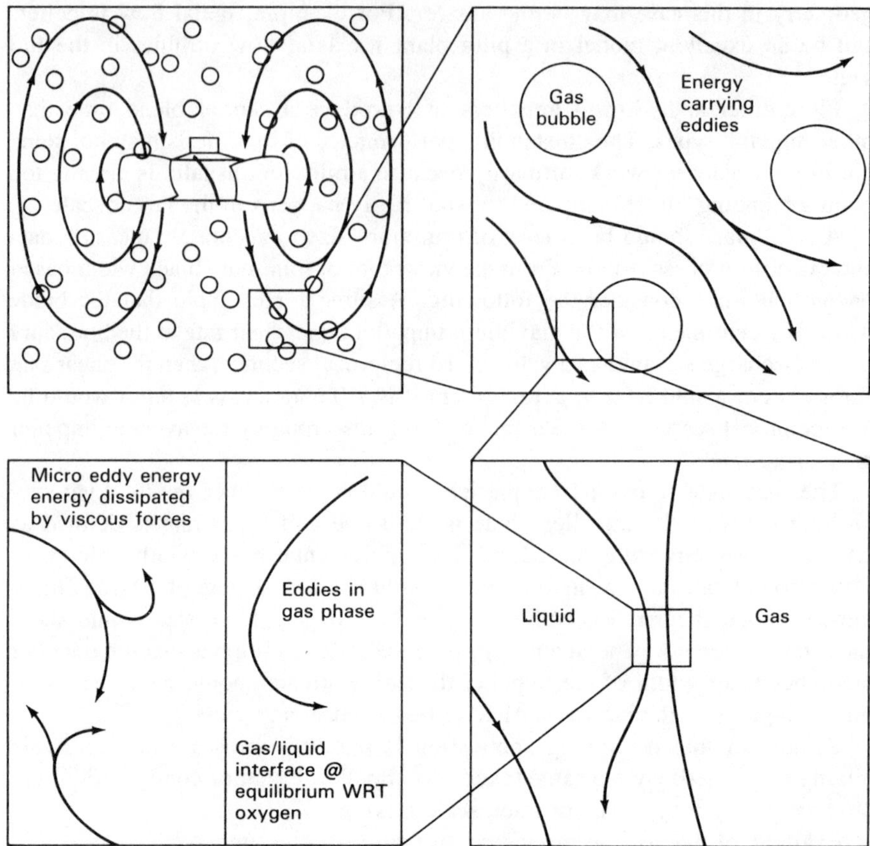

FIGURE 4-8. Macroscale mixing in a fermenter broken down into microscale components. *(Prochem Mixing Equipment, Inc.)*

Momentum theory stipulates that equal bulk fluid velocity and velocities equal at corresponding points in different scale vessels, guarantees similar macroscale mixing results based on scaleup criteria. Macro mixing is deeply rooted in bulk velocity as being the prime scaleup criteria and yardstick to quantify process results.

The average or bulk velocity in an agitated tank is directly proportional to the momentum (QV) of the fluid developed by the impeller. It has been shown that:

$$V_{\text{bulk}} \propto \sqrt{\frac{QV}{(\text{Volume})^{2/3}}} \tag{7}$$

where: V_{bulk} is average of bulk fluid velocity
Q is flow rate
V is velocity
QV is momentum
Volume is volume of fluid in the tank

Equal momentum levels at various scales will yield equal bulk velocities and equal velocities of fluid at corresponding points in different size vessels. Equal velocities in two different scales will produce similar mixing results for macroscale controlled applications.

5

Static Inline Mixing

Static inline mixers do not have a mechanical rotating shaft and impeller assembly acting as the driving force for mixing. Static mixers promote mixing in a pipeline. A pump must provide the driving force to transfer fluids through the inline mixer, overcoming the resistance to flow created by the static mixer elements.

From definition of pipe or plug flow, all particles will have the same residence time in a given length of pipe. This is illustrated in Fig. 5-1.

The implication in plug flow is that there will not be longitudinal or axial mixing and, therefore, time interval uniformity is not possible. In other words, any two cross sections, at any two points in time are not necessarily uniform. However, in any one particular cross section, complete uniformity will exist throughout that section resulting in transverse uniformity. (See Fig. 5-2.)

Static inline mixers are pipeline mixing devices that can provide transverse uniformity. Some practical applications for inline mixers are:

1. Heating a process stream by the addition of steam
2. Blending of intermediate products to obtain a final product
3. pH control incorporating a feedforward and feedback probe to obtain accurate pH control

It must be remembered, however, that to achieve time interval uniformity of mix a stirred vessel is required to provide sufficient residence time in the vessel while the tank contents are being turned over and circulated.

Table 5-1 compares static inline mixing systems and conventional mechanically stirred agitators.

Static mixers have been in use for over 25 years as an alternative to me-

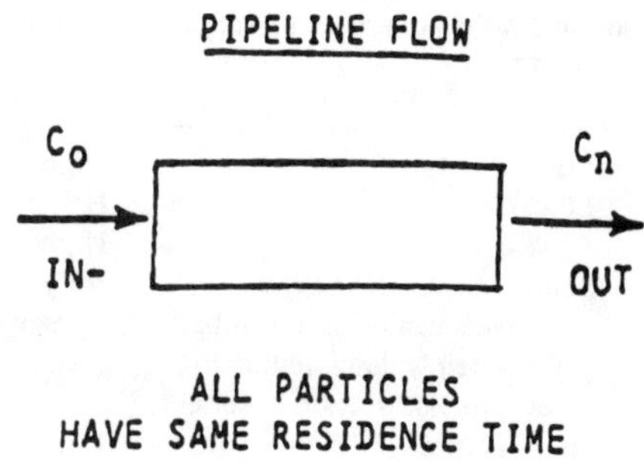

FIGURE 5-1. Turbulent pipe flow or plug flow. *(Mixing Equipment, Inc.)*

chanically stirred agitators. The inline mixer consists of a pipeline housing with stationary elements inside the housing. The precise geometric arrangement of these elements splits the flow of material being pumped through the mixer providing a plug flow regime with minimal backmixing. These elements are designed for either turbulent or laminar flow regimes within the pipeline mixer, and are configured in an alternating pattern such that each element may be 90° out of rotation with the one before and after it in the series. Once again this is to create a progressive splitting and mixing of flow as material moves through the inline mixer housing in a plug flow regime with minimal backmixing.

The mixing elements can be welded edge to edge for removal from the housing in one complete piece or they can be inserted as individual components. Individual elements may be removed (or added if space is provided in the housing) to compensate for changes in process conditions requiring more or less mixing in the static mixer. Additional elements will afford one greater mixing intensity in the static mixer, while fewer elements will result in less mixing. Mixing elements can also be sanitary in design and construction for application

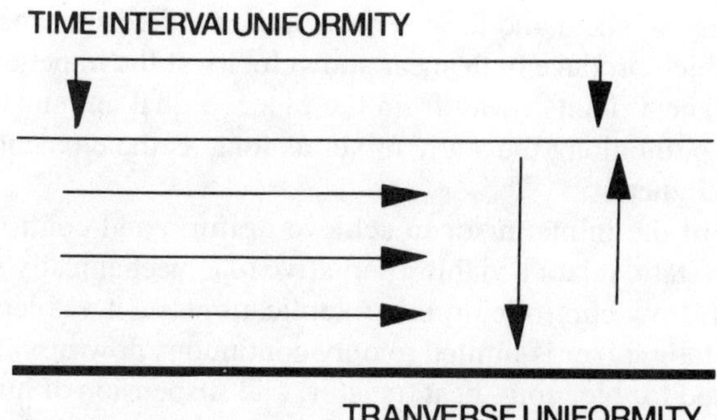

FIGURE 5-2. Transverse versus time interval uniformity. *(Mixing Equipment, Inc.)*

TABLE 5-1. Comparison of Static Mixers and Mechanical Agitators.

	Static Inline	Mechanically Stirred
Capital cost	Lower	Higher
Operating cost	Lower	Higher
Residence time	Low	Higher
Adaptability to process changes	Somewhat Restricted (i.e., elements can be removed but not added without adding a new pipeline section in series)	Very adaptable via speed changes, impeller retrofits, etc.
Potential for plugging up	High (particularly if fluids are "dirty"); will plug up with solids in a power outage	Low (unless power outage "sands in" the impeller)
Flow regime	Plug or pipeline with minimal backmixing; uniform and controlled degree of flow and shear	Transitional or turbulent flow with backmixing; variations in flow and shear throughout
Applicability	Best for blending and heat transfer and hindered settling solid suspension; fair for free settling solid suspension, gas dispersion, and high shear applications; must be continuous flow	Excellent for all applications based on use of proper impeller design; can be batch or continuous flow
Maintenance	Low; no rotating elements to wear out	Needs to be scheduled regularly—oil change, seals, bearings, gears

in the food and pharmaceutical industries. The key to proper element design and configuration is that all material(s) being pumped across them must be rapidly equalized and uniformly distributed such that all concentration gradients are eliminated.

The static mixer is distinguished by its inherent ability to achieve a uniform, controlled degree of shear and flow. This is quite unlike its mechanically stirred counterpart, which produce high shear and velocity at the impeller blade tip and reduced shear and velocity away from the blade. Equal amounts of energy are applied at any point along the static mixer as long as the elements are properly designed and aligned.

The ability of the inline mixer to achieve uniform and controlled shear and flow makes the static mixer a viable alternative to a mechanically stirred agitator for continuous, flow controlled mixing applications such as blending and heat transfer. The static mixer is limited to only continuous flow mixing applications and is best suited for blending, heat transfer, and suspension of hindered settling solids. Free settling solid suspension and higher shear applications (i.e., gas dispersions, mass transfer, extractions, dispersions, etc.) can be accomplished with static mixing, usually less efficiently than with mechanically stirred agitators.

Static mixer housing end connections can be made in various flange styles (i.e., raised or flat face, etc.), weld-prepared, threaded, or with quick-disconnect fittings, for ease of installation and removal.

Injectors and/or sparge assemblies are located on the housing in such a manner to introduce components into the mainstream of flow moving through the mixer. Location and design are critical based on specific volume, density and viscosity ratios.

Inline mixers can be jacketed pipeline units for heat transfer applications. Shell-and-tube designs are engineered for viscous heat transfer requirements. For these situations, elements are placed within the tubes of a shell-and-tube heat exchanger and the viscous fluid is pumped through the tube side. Mixer(tube)-side heat transfer film coefficients are enhanced by maintenance of uniform and controlled velocities along the inside wall of the mixer.

Materials of construction for inline static mixers include steel and high alloys, plastic or plastic coated, rubber and urethane coated, etc. depending on corrosive and/or erosive properties of the fluids being mixed.

A. DESIGNING THE STATIC MIXER

The following is a simple, yet thorough approach to designing a static mixer for specific applications and process results. The graphs and charts illustrate how to select a static mixer for blending or dispersion applications. Both single phase (liquid–liquid, gas–gas) and two phase mixing applications (i.e., gas–liquid) are explained. Methods for estimating liquid blending pressure drops and expected dispersion droplet size are provided. For ease of use, the following presentation takes on a step by step approach and evaluation of various parameters, with appropriate worksheets as noted, to properly design a static inline mixer.

1. Identify the Process

A careful and effective recommendation can only be achieved when all the facts relative to the process are available. As a minimum, the following variables must be determined prior to attempting a selection (other parameters specific to the existing process will be helpful in designing the static mixer):

(a) Components to be mixed
(b) Flow rates of each component, gpm
(c) Existing pipeline sizes, inches
(d) Viscosity of each component, centipoise
(e) Maximum allowable head loss, psi

(f) Specific gravity of each component, SG
(g) Operating temperature and pressure
(h) Materials of construction
(i) Type of inline mixer end connections
(j) Any special requirements

2. Determine Flow Characteristics

The flow regime in the process pipeline must be determined. Whether it is laminar, transitional, or turbulent flow will dictate the preliminary step in sizing an inline static mixer through the correlation of Reynolds number. The calculation of Reynolds number in the pipeline can be made using the following:

$$N_{Re} = \frac{3157\ Q\,\mathrm{SG}}{\mu D} \tag{1}$$

where: N_{Re} = Reynolds number
Q = flow, gpm
SG = specific gravity
μ = viscosity, centipoise
D = pipe diameter, inches.

Mixing requirements in the laminar flow regime will require more mixing elements to provide homogeneity, whereas turbulent flow applications usually require only 2–4 elements.

Therefore, determining the flow regime will not only establish the type of static mixer element (i.e., turbulent or laminar) but also a rough indication of how many.

3. Estimate of the Number of Elements

With the Reynolds number calculated and flow regime established, Fig. 5-3 may be used to determine the minimum number of elements. Emphasis must be placed on proper mixer selection being dependent on other variables. The ultimate selection of the number of static mixing elements may need to be revised in accordance with these factors. Some of the more typical element modifiers are as follows:

(a) Viscosity Ratio. For viscosity ratio between two fluids exceeding 1,000 to 1, add 2 to 6 elements depending on whether the flow regime is turbulent or laminar, respectively.

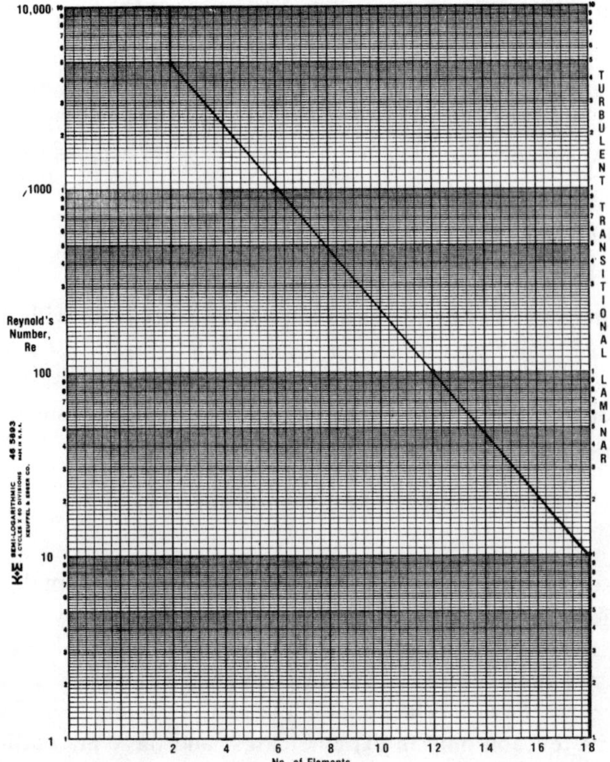

FIGURE 5-3. Selection of number of elements for blending application. *(Eastern Mixers, Inc.)*

(b) Volumetric Flow Rate Ratio. If the volumetric flow rate ratio exceeds 100 to 1, add 2 to 6 elements respectively for turbulent or laminar flow.

(c) Injection Method. An injector is usually mounted on the centerline of the inline static mixer to assure proper injection of a sidestream into the mainstream of flow through the mixer. If the injector is omitted from the design, an additional 1 to 2 elements should be added to allow for "extramixing" needed in lieu of the injector.

(d) Immiscibility and Dispersion. Enhanced phase transfer is the goal when immiscible fluids are mixed. The role of the static mixer is to generate a uniform droplet size. In order to accomplish this, 4 elements are always used except in two special cases. If non-centerline-injection is used, 6 elements are required. If there is a residence time requirement, arising from reaction for instance, use as many elements as are needed.

4. Select Appropriate Diameter and Length—Liquid Blending

Table 5-2 provides a simplified selection of the static mixer diameter and number of elements based on the process flow rate and viscosity. The three options within each box are sized to yield a maximum pressure drop as follows:

Option A—Pressure drop $\leq$ 1 psi
Option B—Pressure drop $\leq$ 5 psi
Option C—Pressure drop $\leq$ 10 psi

These alternatives provide the design/application engineer with the freedom and flexibility of analyzing various pipe sizes and lengths to best suit the piping scheme. Any option will provide the desired mixing and process results. Modifications required per Fig. 5-3 and/or the element modifiers listed in Section 3 may be necessary.

5. Select Static Mixer Diameter and Length for Gas Blending

If various gases must be blended, Table 5-3 can be used as a preliminary estimate for static mixer selection. The blending of gases is very easy for static mixers and generally requires only 2 mixing elements. The table provides anticipated pressure drops based on 2 mixing elements, as well as mixer diameter and gas volumetric flow rate in actual cubic feet per minute, ACFM. The selection requires revision only in extreme cases, and these are outlined below. The table should be used as follows:

(a) Enter the table at the estimated line size and gas flow rate. Choose the final mixer diameter within acceptable pressure drop limits.

(b) Revise the pressure drop for actual density. The chart is based on air with a density of ρ_{air} = 0.076 pound per cubic foot, lb/ft^3. Multiply the table value of pressure drop by the density correction factor, ρ_{actual}/0.076 lb per ft^3.

(c) If the volumetric flow rate ratio of gases is greater than 1,000 to 1 then use 3 elements and multiply pressure drop by 3/2 to compensate for 3 rather than 2 elements.

(d) If the density ratio of gases is greater than 5 to 1, use 4 elements and multiply pressure drop by 4/2 or 2.

(e) If pipe Reynolds number is less than 2,000 (i.e., transitional or laminar flow regime), one should consult a static mixer supplier for selection.

6. Estimate of Pressure Drop Through the Static Mixer

Tables 5-2 and 5-3 provide rough estimates of pressure drops through inline mixers. Fig. 5-4 presents a more accurate assessment of pressure drop. Pressure drop calculation or estimation is an important consideration, since the static

TABLE 5-2. Static Mixer Design and Selections for Liquid Blending Applications.

FLOW (GPM) \ VISCOSITY (cps)		1	10	50	100	500	1000	3000	5000	10K	50K	100K
1	A	½ × 6	½ × 6	½ × 6	¾ × 12	1 × 18	1.5 × 18	2 × 18	2.5 × 18	3 × 18	6 × 18	8 × 18
	B	—	—	—	½ × 12	¾ × 18	1 × 18	1.5 × 18	1.5 × 18	2 × 18	3 × 18	4 × 18
	C	—	—	—	—	½ × 12	¾ × 18	1 × 18	1 × 18	1.5 × 18	2.5 × 18	3 × 18
5	A	¾ × 6	¾ × 6	1 × 6	1.5 × 12	2 × 12	2.5 × 18	4 × 18	4 × 18	6 × 18	8 × 18	10 × 18
	B	½ × 6	½ × 6	¾ × 6	¾ × 6	1.5 × 12	1.5 × 18	2.5 × 18	3 × 18	3 × 18	6 × 18	8 × 18
	C	—	—	½ × 6	½ × 6	1 × 12	1 × 12	2 × 18	2 × 18	2.5 × 18	4 × 18	6 × 18
10	A	1.5 × 6	1.5 × 6	1.5 × 6	2 × 6	2.5 × 12	3 × 12	4 × 18	6 × 18	8 × 18	10 × 18	12 × 18
	B	1 × 6	1 × 6	1 × 6	1 × 6	1.5 × 12	2 × 12	3 × 18	3 × 18	4 × 18	8 × 18	10 × 18
	C	¾ × 6	¾ × 6	¾ × 6	¾ × 6	1 × 12	1.5 × 12	2.5 × 18	3 × 18	3 × 18	6 × 18	6 × 18
25	A	2 × 6	2 × 6	2 × 6	2.5 × 6	3 × 12	4 × 12	6 × 18	8 × 18	10 × 18	12 × 18	
	B	1.5 × 6	1.5 × 6	1.5 × 6	1.5 × 6	2 × 12	2.5 × 12	4 × 18	6 × 18	6 × 18	10 × 18	
	C	1 × 6	1 × 6	1 × 6	1 × 6	1.5 × 12	2 × 12	2.5 × 12	3 × 18	4 × 18	8 × 18	
50	A	2.5 × 4	2.5 × 4	2.5 × 4	3 × 6	4 × 12	6 × 12	8 × 18	10 × 18	12 × 18		
	B	2 × 6	2 × 6	2 × 6	2 × 6	2 × 6	3 × 12	4 × 12	6 × 18	8 × 18		
	C	1.5 × 6	1.5 × 6	1.5 × 6	1.5 × 6	2 × 6	2.5 × 12	3 × 12	4 × 18	6 × 18		
100	A	2.5 × 2	3 × 2	4 × 4	4 × 6	6 × 6	8 × 12	10 × 12	10 × 18	12 × 18		
	B	2.5 × 2	2.5 × 2	2.5 × 4	2.5 × 4	3 × 6	4 × 12	6 × 12	6 × 12	10 × 18		
	C	2 × 6	2 × 6	2 × 6	2 × 6	2.5 × 6	3 × 6	4 × 12	6 × 12	8 × 18		
200	A	4 × 2	4 × 2	6 × 4	6 × 4	6 × 6	8 × 12	10 × 12	12 × 12			
	B	3 × 2	3 × 2	4 × 4	4 × 4	4 × 6	6 × 6	8 × 12	8 × 12			
	C	2.5 × 2	2.5 × 2	2.5 × 2	3 × 4	3 × 4	4 × 6	6 × 12	6 × 12			
600	A	6 × 2	6 × 2	8 × 4	8 × 4	10 × 6	12 × 6					
	B	4 × 2	4 × 2	6 × 2	6 × 4	8 × 6	8 × 6					
	C	4 × 2	4 × 2	4 × 2	4 × 2	6 × 6	6 × 6					
1000	A	8 × 2	8 × 2	8 × 2	10 × 4	12 × 6						
	B	6 × 2	6 × 2	6 × 2	6 × 4	10 × 6						
	C	6 × 2	6 × 2	6 × 2	6 × 4	8 × 6						
2000	A	10 × 2	10 × 2	12 × 2	12 × 2							
	B	8 × 2	8 × 2	8 × 2	8 × 2							
	C	6 × 2	6 × 2	8 × 2	8 × 2							
4000	A	18 × 2	18 × 2	20 × 2								
	B	12 × 2	12 × 2	12 × 2								
	C	10 × 2	10 × 2	10 × 2								
6000	A	24 × 2	24 × 2									
	B	14 × 2	14 × 2									
	C	12 × 2	12 × 2									

LEGEND

A	6 × 6
B	4 × 6
C	3 × 4

A — Option for 1 psi
B — Option for 5 psi
C — Option for 10 psi
6 × 6: Mixer Diameter (inches) × No. of Elements

Note Selections based on S.G. = 1.0

TABLE 5-3. Static Mixer Design and Selections for Gas Blending Applications.

2″		4″		6″		8″	
ACFM	ΔP	ACFM	ΔP	ACFM	ΔP	ACFM	ΔP
100	0.23	100	0.01	500	0.06	500	0.01
200	0.94	200	0.05	1000	0.22	1000	0.07
300	2.14	300	0.11	1500	0.51	1500	0.15
400	3.85	400	0.19	2000	0.87	2000	0.26
500	6.08	5000	0.30	3000	2.00	3000	0.60
700	12.18	700	0.59	4000	3.60	4000	1.07
1000	25.51	1000	1.22	5000	5.70	5000	1.69

10″		12″		14″		16″	
ACFM	ΔP	ACFM	ΔP	ACFM	ΔP	ACFM	ΔP
1000	0.03	1000	0.01	3000	0.09	3000	0.04
2000	0.11	2000	0.05	5000	0.24	5000	0.09
3000	0.24	3000	0.10	8000	0.61	8000	0.24
5000	0.66	5000	0.29	10000	0.95	10000	0.37
8000	1.72	8000	0.74	15000	2.15	15000	0.82
10000	2.72	10000	1.17	20000	3.85	20000	1.47
15000	6.27	15000	2.69	30000	8.79	30000	3.33

24″		30″		36″		42″	
ACFM	ΔP	ACFM	ΔP	ACFM	ΔP	ACFM	ΔP
5000	0.38	15000	0.14	25000	0.19	25000	0.10
20000	0.65	20000	0.25	40000	0.25	40000	0.25
30000	1.50	30000	0.57	60000	1.06	60000	0.55
40000	2.69	40000	1.01	80000	1.89	8000	0.98
50000	4.22	50000	1.59	100000	2.97	100000	1.54
60000	6.12	60000	2.29	125000	4.66	125000	2.42
70000	8.38	70000	3.14	150000	6.72	150000	3.48

mixer is a resistance to flow of fluids (liquid or gases) through it. The driving force of the pump, or blower/compressor for gases, must overcome this added resistance or head in the system in delivering the required flow rate to and through the static mixer.

The graph in Fig. 5-4 is based on water at a specific gravity, SG = 1.0, and viscosity of 1.0 centipoise. Enter the graph at the process flow rate in gpm and at the appropriate mixer diameter. It should be noted that pressure drop is represented as psi per element. Conversion of psi to feet of water is obtained by multiplying the value in psi by 2.303. For those situations where the viscosity or specific gravity exceeds waterlike conditions, a manual determination of pressure drop can be made. Refer to Step 10.

7. Check Velocity Requirements

Line velocity is not usually a parameter used for sizing and selecting a static mixer. However, certain mixing applications require a minimum velocity. The

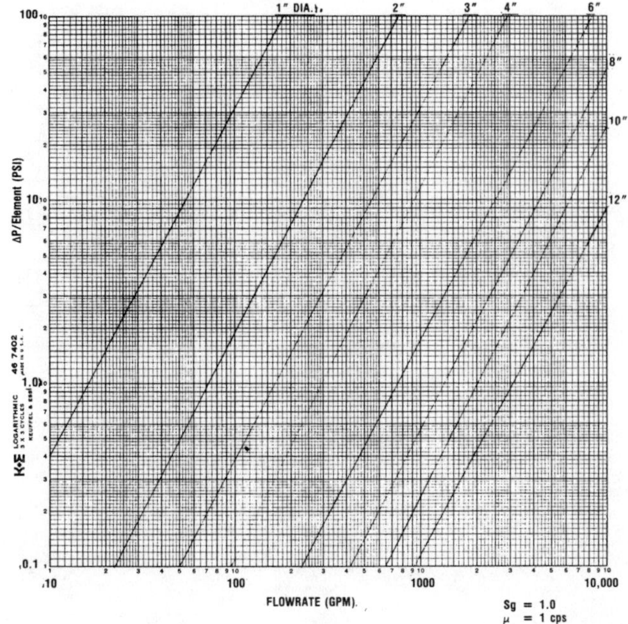

FIGURE 5-4. Pressure drop per element versus flow rate for various diameter static mixers. *(Eastern Mixers, Inc.)*

graph in Fig. 5-5 provides a quick estimate of linear velocity based on mixer diameter and flow rate. The graph should be used in conjunction with the minimum velocity table listed below, to insure that adequate velocities are maintained, where the particular process dictates.

In addition, Fig. 5-5 provides a simple scaleup criterion based on velocity. A static mixer can be scaled up from a pilot unit by observing a combination of several variables, including Reynolds number, linear velocity, shear rate, energy per mass, pressure drop and residence time (for time dependent reactions). Linear fluid velocity is chosen as the constant scaleup factor due to the simplicity of maintaining geometric similarity. It must be noted again, however, that its use is limited to simple blending applications and/or those requirements where velocity is the key design variable. Some of these velocity dependent applications are presented in Table 5-4.

8. Estimate of Drop Size for Liquid–Liquid Dispersions

As discussed previously, dispersion applications require 4 mixing elements. The graph in Fig. 5-6 illustrates the mean drop size to be expected with a prop-

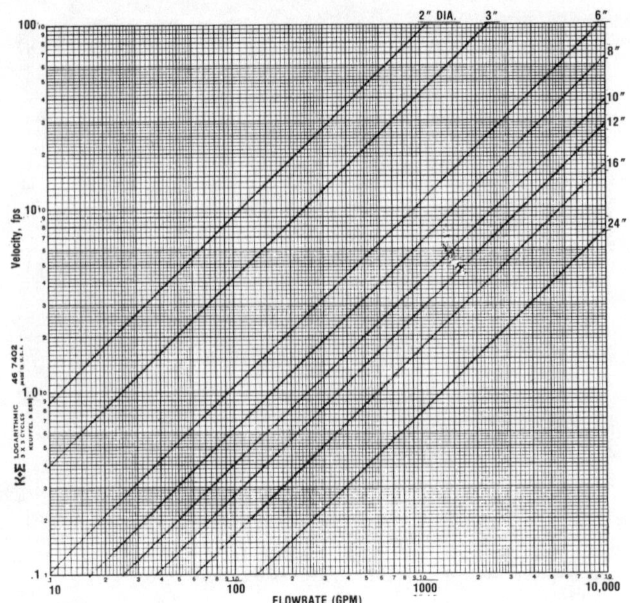

FIGURE 5-5. Mean bulk fluid velocity as a function of flow rate and static mixer diameter. *(Eastern Mixers, Inc.)*

erly designed static mixer. For liquid–liquid dispersions the viscosity ratio must be less than 5 to 1. If the viscosity ratio exceeds 5 to 1, the average drop size must be modified. If the interfacial surface tension is different from that used in Fig. 5-6, i.e., 30 dynes per centimeter, the drop size can be recalculated based on:

$$\overline{D} = 1.14 \text{ We}^{-3/4} D \tag{2}$$

where: D = Sauter mean drop size, inches
D = Pipe diameter, inches
We = Weber number

$$\text{We} = \frac{394\, Q^2 \text{SG}}{\sigma D} \tag{3}$$

where: Q = flow rate, gpm
SG = specific gravity
σ = interfacial surface tension, dynes per centimeter

TABLE 5-4. Velocity Dependent Mixing Applications for Static-Mixers.

Process	Recommended Minimum Velocity Through the Static Mixer[a]	Recommended Minimum Number of Elements
Liquid-Liquid Dispersions[b]		
Drop size = 1,000 microns	2 ft/sec	4
100 microns	5 ft/sec	4
10 microns	7–8 ft/sec	4
Liquid emulsions	10 ft/sec	4
Multiphase dispersion	5–10 ft/sec	4
Dissolved air separation	10 ft/sec	4
Flash mixing (waste water treatment)	3 ft/sec	2
Chlorine bleaching of paper pulp	2–3 ft/sec	8
Simple blending applications	1–5 ft/sec	2

[a]It is recommended that a minimum fluid velocity of 1.0 ft/sec be maintained for turbulent flow applications.
[b]Exact values for liquid-liquid and gas-liquid drop size are presented in Figures 5-6 and 5-7.

9. Estimate of Drop Size for Gas–Liquid Dispersions

For gas–liquid dispersion applications, the viscosity of the liquid must always be less than 10 centipoise to insure the accuracy of the mean drop size predicted in the graph in Fig. 5-7. Viscous forces override the interfacial surface tension above 10 centipoise viscosity, and erroneous results will be estimated using Fig. 5-7.

The graph is based on a 50/50 volumetric flow ratio. Less than 50% gas will only affect the pressure drop. However, if the gas volume exceeds 50%, the entire flow regime of the gas–liquid mix becomes extremely complex and must be studied in detail, preferably in a pilot plant program.

If the interfacial surface tension is other than 80 dynes per centimeter, the drop size must be recalculated as follows:

$$\overline{D} = 0.390 \text{ We}^{-0.43} D \qquad (4)$$

where: D = Sauter mean drop size, inches
We = Weber number (see Equation 5-3)
D = Pipe diameter, inches

10. Manual Calculation of Pressure Drop

Three criteria are required to determine the pressure drop across a static mixer:

(a) Reynolds number, N_{Re}
(b) Pressure drop in empty pipe of same diameter as inline mixer

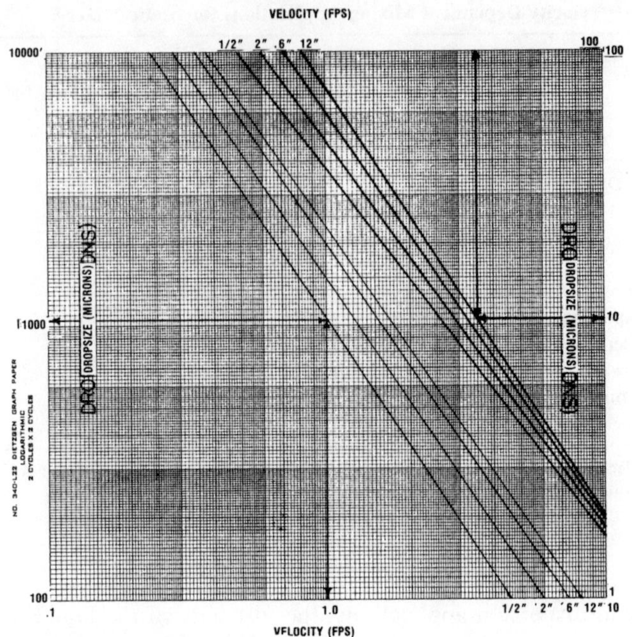

Figure 5-6. Mean drop size for liquid–liquid dispersions as a function of velocity and mixer diameter. *(Eastern Mixers, Inc.)*

(c) Pressure drop in the static mixer by using the appropriate value of flow coefficient C_f

(a) Reynolds number, N_{Re}:

$$N_{Re} = \frac{3157Q \text{ SG}}{\mu D} \tag{5}$$

where: Q = flow rate, gpm
SG = specific gravity
D = inside diameter of static mixer, inches
μ = absolute viscosity, centipoise

(b) Pressure drop of empty pipe:

$$\Delta P_{pipe} = 1.35 \times 10^{-2} \frac{f \times L \times \text{SG} \times Q^2}{D^5} \tag{6}$$

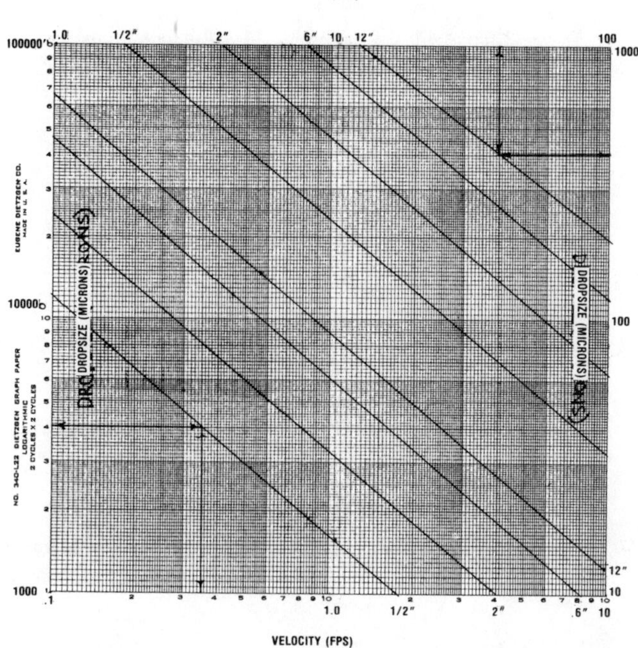

FIGURE 5-7. Mean drop size for gas–liquid dispersions as a function of velocity and mixer diameter. *(Eastern Mixers, Inc.)*

where: ΔP_{pipe} = Pressure drop in empty pipe, psi
f = Darcy friction factor (see Fig. 5-8)
L = Length of pipe, feet
D = Pipe diameter, inches
SG = Specific gravity
Q = Flow rate, gpm

(c) Pressure drop in the static mixer by using C_f. The flow coefficient C_f is a calculated factor to compensate for head loss through the static mixing elements. For N_{Re} less than 10, $C_f = 6$. For $10 < N_{\text{Re}} < 1000$,

$$C_f = 1.53(N_{\text{Re}})^{0.45} \tag{7}$$

For $N_{\text{Re}} > 1000$,

$$C_f = -15.9 + 8.41 \ln N_{\text{Re}} \tag{8}$$

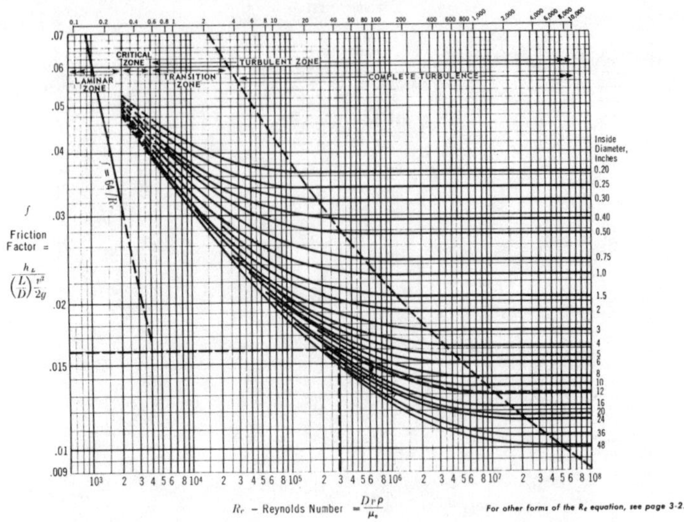

FIGURE 5-8. Darcy friction factors for clean commercial steel and wrought iron pipe. *(Eastern Mixers, Inc.)*

The overall pressure drop through the static mixer may be expressed as:

$$\Delta P = \Delta P_{\text{pipe}} \times C_f \tag{9}$$

where: ΔP = Pressure drop through the mixer, psi
ΔP_{pipe} = Pressure drop through empty pipe, psi
C_f = Flow coefficient per eqs. (7) and (8)

11. Estimate of *G* Value for Water Treatment Applications

The *G* value correlation has been very prominent as a *relative* measure of mixing for municipal water treatment mixing applications, such as flash mixing, flocculation, conditioning, etc. While generally used in selecting mechanically stirred agitators, the following equation is applicable for static mixers as well, based on water at 20° centigrade.

$$G = 188.6 \sqrt{\frac{f \times V^3 \times C_f}{D}} \tag{10}$$

where: G = ft/sec per foot of mixer length
f = Darcy friction factor (see Fig. 5-8)
V = Fluid velocity, ft/sec
D = Inside diameter of pipe, ft
C_f = Flow coefficient

Flash mixers will normally require G values between 500 to 1000 feet per second linear foot of static mixer.

6

Retrofitting: A Cost Effective Way to Convert Yesterday's Mixer to Today's Needs

A. RETROFITTING OF IMPELLERS

If a particular mixer, equipped with traditional axial flow or radial flow turbines, is not achieving desired mixing results, or is achieving desired mixing results but at excessive power requirements, it is a candidate for a retrofit high-efficiency hydrofoil impeller. In either case, replacing the energy intensive, conventional turbines with energy efficient, state-of-the-art hydrofoils will most likely resolve either problem.

It is very often not necessary to replace an entire agitator with a new one to rectify these situations. This is an expensive solution. A more cost-effective scenario is to consider replacing just the impellers while keeping the balance of the agitator intact (the same drive, motor, shaft). The retrofitted hydrofoil impellers would be larger in diameter than the traditional ones being replaced. This is because of two factors:

1. Hydrofoil impellers have lower power numbers than traditional axial or radial flow impellers (i.e., fixed pitch flat-bladed turbines, propellers, radial flow disk turbines and paddles, etc.) The same power is required from the hydrofoil even though its diameter is larger.

2. Impeller speed is usually kept constant. Since flow Q is proportional to ND^3,

$$Q \propto ND^3 \tag{1}$$

where: N = impeller speed
D = impeller diameter,

by retrofitting at constant power and speed, in such a manner that the impeller diameter *D* of a hydrofoil increases, flow will increase commensurately.

Flow controlled mixing applications such as blending, solid suspension and heat transfer will benefit the most by retrofitting with hydrofoils. If a specific mixer with flat-bladed axial or radial flow impellers or propeller(s) is not accomplishing desired mixing results (for example blend time is too long or suspension of solids is inadequate), or is performing well but at the expense of too much power, improvements in either case may be had by replacing the impellers with hydrofoils.

B. FERMENTATION RETROFITS

Shear intensive mixing applications, with a flow or bulk fluid velocity requirement, such as gas–liquid mixing applications in aerobic fermentation, can be optimized with retrofitted hydrofoils. It has been demonstrated that mass transfer rates can be increased at constant power, to enhance yield in bacterial and mycellial fermenters, when radial flow Rushton type turbines are replaced/retrofitted with hydrofoils. In addition, heat transfer rates are increased due to greater bulk fluid velocity across the heat transfer surface area. Assuming mass transfer rates are adequate, significant power savings may be realized by replacing Rushton radial flow in impellers with hydrofoils. Power savings can be impressive with large commercial fermenters with installed horsepowers typically ranging from 10–25 horsepower per 1,000 gallons.

Much of the interest in hydrofoil impellers for gas–liquid processes come from studies directed at retrofitting existing fermentation tanks. These tanks typically had power levels from 10 to 25 horsepower per 1,000 gallons. Most often, multiple radial flow Rushton impellers on a mixer were used. Superficial gas velocities were on the order of 0.1 to 0.5 feet per second.

A more efficient blending impeller gives better results when a sufficient or excess gas–liquid mass transfer ability is provided by the existing mixer and air rate. A nonuniform oxygen concentration throughout the tank can exist because of poor blending. However, some damage may occur to the organism from the high input energy level going into macro- and microscale shear rates from the radial flow impellers. Retrofitting the radial flow impellers with hydrofoils may minimize shear related damage to organisms, particularly at reduced power levels.

A test program is recommended to determine if all the radial flow turbines can be retrofitted with hydrofoils—very often the lower impeller must remain as a radial flow turbine to ensure uniform dispersion of the gas. Also wide- and narrow-bladed hydrofoils should be investigated to determine the ultimate combination/configuration of Rushton impellers, wide and narrow-bladed hydrofoils, producing optimum process results.

Wide-bladed hydrofoils with solidity ratio (i.e., ratio of actual blade area to the circular area swept by the blades) of at least 85% will disperse gas without flooding at superficial gas velocities typically encountered in a fermenter (i.e., 0.1 to 0.5 feet per second). The mass transfer coefficients versus mixer power level at various gas rates are comparatively greater with wide-bladed hydrofoils than radial flow Rushton impellers.

Experimentation with wide-bladed hydrofoils in a comparative analysis with Rushton turbines in fermenters has indicated the following:

(a) The slope of the mass transfer coefficient as a function of power and gas rate curves has a marked breakpoint or "knee" in the plot for a Rushton turbine. This characteristic is not observed with wide-bladed hydrofoils. Process results, particularly on scaleup, are therefore more predictable with wide-bladed hydrofoils.

(b) Retrofitting existing mixers is common. Typically this involves keeping mixer speed and power the same while changing impeller geometry and size (from Rushtons to larger diameter hydrofoils). As a result, one is maintaining constant torque on the equipment. The gearbox speed reducer does not have to be changed or modified.

(c) When the ratio of impeller to tank diameter, D/T, for a radial flow Rushton turbine is on the order of 0.33 to 0.4, the gas–liquid mass transfer characteristics of wide-bladed hydrofoils will be comparatively better than those of the Rushton turbines. Mass transfer, overall blending, and heat transfer will be improved.

(d) Testing must be conducted with various combinations and configurations of wide-bladed and narrow-bladed hydrofoils and Rushton turbines to optimize process results and establish scaleup criteria.

(e) Shear sensitive fermentations are best served by hydrofoils. Narrow- or wide-bladed geometries and configurations are predicated by the shear/flow balance commensurate with required process results.

Wide-bladed hydrofoils increase mass transfer coefficients, decrease shear rates, improve blending and heat transfer and reduce power consumption when they replace traditional radial flow Rushton turbines in fermenters. When retrofitted at equal power consumption and speed, wide-bladed hydrofoils can increase mass transfer coefficient by 25–40%.

Productivity increase may not be the justification for retrofitting. Horsepower requirements can be reduced by 30–50%, at equal productivity, by replacing Rushton turbines with hydrofoils in fermentation services. This reduction in power can be particularly significant in large fermenters, often equipped with 200–500 horsepower agitators.

Hydrofoils generate one-fourth the shear rate of shear intensive Rushton tur-

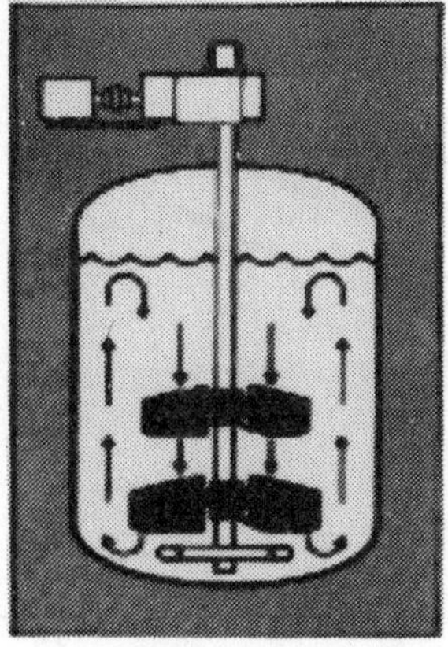

FIGURE 6-1. Sectional view of a fermenter with wide bladed hydrofoils. *(Mixing Equipment, Inc.)*

bines, when retrofitted at equal power and speed. Minimum turbulence and maximum fluid flow make hydrofoils ideal for applications where lower maximum macroscale fluid shear levels are beneficial. Shear sensitive mammalian cell fermentations are best served by hydrofoils.

Figures 6-1 and 6-2 represent sectional views of fermenters equipped with wide-bladed hydrofoils and Rushton turbines, respectively. For a retrofit campaign going from Rushtons to hydrofoils, (i.e., Fig. 6-2 to 6-1), the mixer power and operating speed are usually kept constant (i.e., there is no change to the drive assembly—gear box and motor). The mixer shaft is usually un-

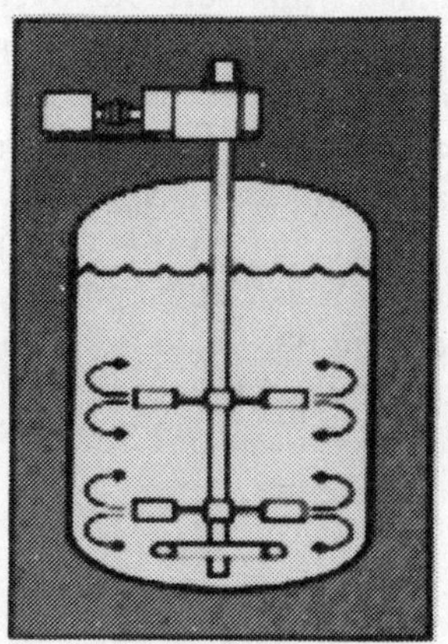

FIGURE 6-2. Sectional view of a fermenter with Rushton turbines. *(Mixing Equipment, Inc.)*

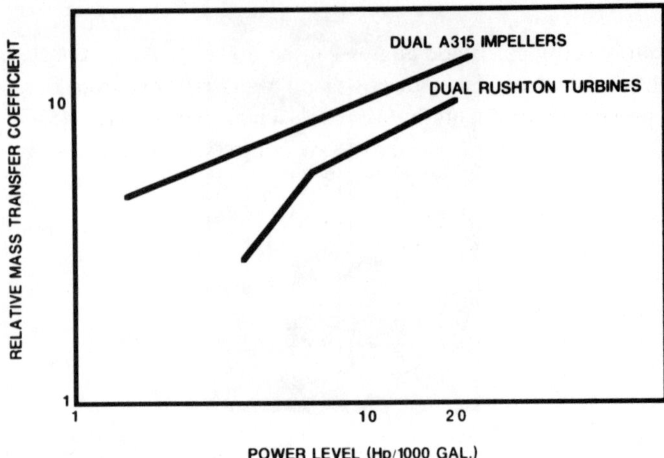

FIGURE 6-3. Mass transfer coefficient as a function of power level and impeller geometry in fermenters. *(Mixing Equipment, Inc.)*

changed, unless a mechanical check indicates existing shaft is incapable of withstanding the higher mechanical loads. Larger diameter hydrofoil impellers replace smaller Rushton turbines. Tank internals are not changed in a traditional retrofit scenario.

Figure 6-3 shows the incremental increase in mass transfer coefficient of the hydrofoils over Rushtons in retrofitted fermenters at various power levels and constant speed. One can see the comparative improvement in mass transfer at any given power level, when hydrofoils retrofit Rushton in fermenters. In a similar manner, the relative decrease in power requirement at any given mass transfer coefficient, is illustrated with hydrofoils over Rushtons.

The data needed to generate Fig. 6-3 can be obtained in any scale fermenter equipped with a variable speed agitator and mass transfer instrumentation, such as a dissolved oxygen meter and discharge gas analysis equipment.

The merits of Fig. 6-3 have been discussed from a scaleup perspective. With the relationship of mass transfer and power level plotted as shown, proper scaleup factors may be applied to assure equal process results in the larger scale.

Another reason for developing the relationship between mass transfer and power level for Rushton turbines and hydrofoils, as illustrated in Fig. 6-3, is to predict the results of improved mass transfer or reduced power level when retrofitting hydrofoils into a fermenter.

C. FLOW CONTROLLED RETROFITS

Low viscosity, flow controlled applications such as blending, heat transfer, and solids suspension, are best suited for retrofitted hydrofoils with narrow blades.

Very often the existing mixer will be equipped with conventional, constant-pitch, axial flow turbines. Blade angles can be 30–45°, but of constant pitch.

Narrow-bladed hydrofoils, with power numbers ranging from 0.15 to 0.30, can reduce power requirements by 40–70% when they replace 45° axial flow turbines, while still achieving the same flow controlled mixing results. Thus, if an existing agitator with conventional fixed pitch axial turbines is achieving satisfactory mixing results, albeit at high power usage, considerable power savings may be had by retrofitting with narrow-bladed hydrofoils without jeopardizing process results.

Hydrofoils will increase flow incrementally by 70–150% over 45° axial flow turbines, in a retrofit situation at equal speed and power. Thus, significant improvement in pumping capacity is seen with narrow-bladed hydrofoils, with power numbers ranging from 0.15 to 0.30, retrofitted in low viscosity (less than 3,000 centipoise) flow controlled applications. Needless to say, an improvement in flow of this magnitude is very often enough to turn an unsatisfactory mixer into one achieving satisfactory results.

The costs associated with making the retrofit (i.e., capital cost and installation) are more than offset by lower energy consumption and/or increased productivity, as well as reduced mechanical wear in the drive assembly.

Agitators can be a major factor in plant power consumption, thus paving the way for potential retrofitting. Higher efficiency hydrofoils mean reduced power costs for any given mixing job. Hydrofoils that have replaced conventional axial flow turbines can accomplish the same mixing results for as little as one-third the power requirements.

Table 6-1 illustrates comparative power requirements and savings with retrofit hydrofoils versus conventional 45° axial flow turbines. Payback of the capital and installation costs associated with the hydrofoil is often measured in months. The table is based on high efficient, 4-bladed hydrofoils with power number of 0.15, achieving similar flow controlled mixing results as that with 45° axial flow turbines.

TABLE 6-1. Power Savings of Narrow-Bladed Hydrofoil versus 45° Axial Flow Turbine at Equal Mixing Results.

Axial Turbine Hp	Rotofoil Hp	Annual $ Savings at 8¢ per Kwh
5	1.5	$ 1830
10	3	$ 3660
15	5	$ 5228
25	7.5	$ 9150
30	10	$10,457
50	15	$18,300
75	20	$28,757

The life of mixer gearbox and the rotating elements of the agitator are a function of the impeller generated torque. As the torque decreases, bearing life increases to the cubed power or,

$$BL \propto 1/(\text{Torque})^3 \tag{2}$$

where BL = bearing, life. For example, reducing applied torque to a gearbox by 50% increases bearing life by 800%. A hydrofoil used to replace a typical axial flow, constant pitch turbine can reduce gearbox torque by as much as 70%, while still achieving similar mixing results. Torque reduction is directly proportional to power reduction since speed is held constant during a retrofit situation.

$$\text{Torque} \propto \text{Hp}/N \tag{3}$$

where: Hp = Power
N = Speed.

Table 6-2 presents comparative torque reductions and increased gearbox service factors for retrofit hydrofoils, replacing 45° axial flow turbines, while achieving similar flow controlled mixing results. Operating speed of the mixers has been held constant in all cases at 100 rpm. Higher efficiency with the hydrofoil means lower torque with greatly extended gearbox life—without jeopardizing process results.

Impeller pumping capacity is directly related to tank mixing efficiency. Inefficient mixing means a longer production cycle, therefore higher production costs. An existing agitator can be retrofitted to create as much as 70 to 150% greater incremental pumping capacity at equal horsepower, if the existing axial flow constant pitch turbine is replaced with a flow efficient hydrofoil. Low vis-

TABLE 6-2. Torque Reduction with Hydrofoils versus 45° Axial Flow Turbines at Equal Mixing Results.

Axial Turbine Hp	Rotofoil Hp	Axial Turbine Torque	Rotofoil Torque	Increased Gearbox Service Factor
5	1.5	3151	945	333%
10	3	6302	1891	333%
15	5	9454	3151	300%
25	7.5	15,756	4727	333%
30	10	18,907	6302	300%
50	15	31,513	9454	333%
75	20	47,269	12,605	375%

cosity flow controlled mixing applications (less than 3,000 centipoise bulk viscosity) are optimized by retrofitting with narrow width, three or four bladed, hydrofoils with power numbers of 0.15 to 0.30. Higher viscosity flow controlled mixing applications are optimized by retrofitting with wide-bladed (solidity ratios of at least 85%), three or four bladed hydrofoils. Realistically, wide-bladed hydrofoils are cost effective retrofits in a rather limited envelope of viscosity, typically from 3,000 to 10,000 centipoise. Above 10,000 centipoise, hydrofoils tend to become inefficient in producing adequate flow against higher viscous drag resistances. Alternative impeller choices should be sought for optimized retrofitting. These include 45° axial flow turbines, anchors and helical impellers.

A traditional mixer with a constant pitch axial flow turbine can become a ''state-of-the-art'' production machine by retrofitting with hydrofoils at a fraction of the cost of a new mixer. Low power number hydrofoils afford as much as a 60% increase in impeller diameter for a given power level and constant speed. Since pumping capacity is proportional to impeller diameter cubed (i.e., $Q \propto D^3$) at constant speed, impeller flow will increase exponentially.

Table 6-3 reflects the degree of increased pumping capacity by retrofitting from conventional axial flow turbines to hydrofoils, at equal power and speed.

Another reason to consider the merits of retrofitting with a hydrofoil is to extend the life expectancy of the impeller in an erosive mixing environment. For example, assume a higher shear impeller like a constant pitch axial flow turbine, radial flow Rushton or paddle turbine, is experiencing erosive wear on the blades. Retrofitting these impellers with a lower shear hydrofoil will minimize blade wear without jeopardizing process results. In fact, process results are likely to be improved while at the same time improving the wear characteristics of the impeller blades. Even though tip speed will increase with the retrofitted hydrofoil because of increased diameter at constant speed, maximum shear rate will decrease incrementally by as much as 75% due to the foil blade

TABLE 6-3. Flow Increase with Hydrofoils versus 45° Axial Flow Turbines at Equal Power and Speed.

Present Mixer Hp	Present Axial Flow Turbine Pumping Rate	Hydrofoil Replacement Pumping Rate	Hydrofoil's Incremental Increase in Pumping Rate
5	12,515 gpm	30,432 gpm	143%
10	20,958 gpm	46,750 gpm	123%
15	24,444 gpm	62,224 gpm	154%
25	37,177 gpm	87,704 gpm	136%
30	42,240 gpm	95,010 gpm	125%
50	53,704 gpm	128,282 gpm	139%
75	67,075 gpm	157,781 gpm	135%

shape and uniform velocity distribution and profile. Therefore, lower fluid shear rates result in less blade turbulence and consequently prolonged life expectancy of hydrofoil blades in erosive mixing environments (i.e., erosive slurries).

D. MECHANICAL CONSIDERATIONS OF RETROFITTING

Retrofitting with hydrofoil impellers to improve mixing results and/or reduce power consumption must have a mechanical check to ensure the integrity of the mixer components. Hydrofoils create greater mechanical loads than conventional turbines. Care must be taken (in retrofitting to improve process results or optimizing energy usage) that mechanical problems do not surface.

A detailed discussion of mechanical design of agitators will follow in the next chapter. It is important to know the mechanical parameters that must be considered when contemplating a retrofit. These mechanical considerations will be identified in this chapter. The calculations to quantify these mechanical factors will be addressed in the next chapter.

The usual premise in retrofitting is to keep the existing mixer components the same, except for the impeller. Usually the impeller is replaced with a hydrofoil of larger diameter. The hydrofoil impeller hub and blade assembly is new, while all other mixer components (i.e., shafts, drive assembly, motor) remain as is.

If the hydrofoil is narrow-bladed with three blades, it will weigh less than conventional fixed pitch axial turbines with four or more blades and radial flow turbines of four or more blades, even though the hydrofoil impeller is of larger diameter. Furthermore, hydrofoils can be fabricated of composite materials to considerably lessen the weight of the impeller. Reduced impeller weight will increase the shaft critical speed proportionally to the square root of the ratio of heavier to lighter weight. Therefore, retrofitting with a lighter weight hydrofoil will increase the shaft critical speed. Since operating speed is kept constant, the ratio of operating to critical speed decreases when retrofitting with lighter impellers. Shaft failure will not occur when retrofitting with lighter weight impellers due to critical shaft speed considerations. This is a positive in assuring mechanical integrity of shaft design when considering a retrofit situation. Critical speed inherently increases with lighter weight impellers and hence, if operating speed remains constant, the mixer will be running farther away from critical.

Because of the high flow efficiency of hydrofoils, fluid thrust is greater with hydrofoils than with traditional axial or radial flow turbines. Thrust is proportional to fluid momentum or (flow × velocity). If power is kept constant, thrust will increase about 40% when a hydrofoil replaces a conventional fixed pitch axial or radial flow impeller, due to increased bulk fluid velocity and momentum. Downpumping hydrofoils will create an upward thrust force and vice versa for uppumping impellers. The increased force due to greater thrust must be taken into account with downward forces due to the impeller weight in evaluating the net force and direction acting along the mixer shaft. Fig. 6-4 illustrates thrust force and impeller weight loads in a retrofitted downpumping scenario.

As previously discussed retrofitted hydrofoils weigh less and provide greater thrust than conventional impellers at equal or less power consumption. One must evaluate the overall net effect, which will be a net increase in upward load or force along the shafting into the drive assembly. The greater axial load imposed on the mixer mounting fixture and bearings and gears in the drive assembly and mechanical seal cartridge (if there is a mechanical seal), must be quantified and checked against allowable loads by the original equipment manufacturer (i.e., the original mixer supplier). Axial loads exceeding allowable limits could result in premature bearing or gear failure in the drive assembly, bearing failure in the mechanical seal housing, or possible bolt failure in the mixer mounting plate or flange.

Fluid forces created by a hydrofoil impeller are greater than those of conventional turbines. These forces act on the blades of the impeller and are reactions imparted by the tank surroundings. Flow emanating from the impeller is converted into a force acting back on the blades as flow recirculates and is diverted by tank walls and baffles toward the impeller. The hydraulic load due to fluid force is proportional to a turbine force coefficient and impeller diameter rasied to the 3.5 power (i.e., $D^{3.5}$). Hydrofoil turbine force coefficients are

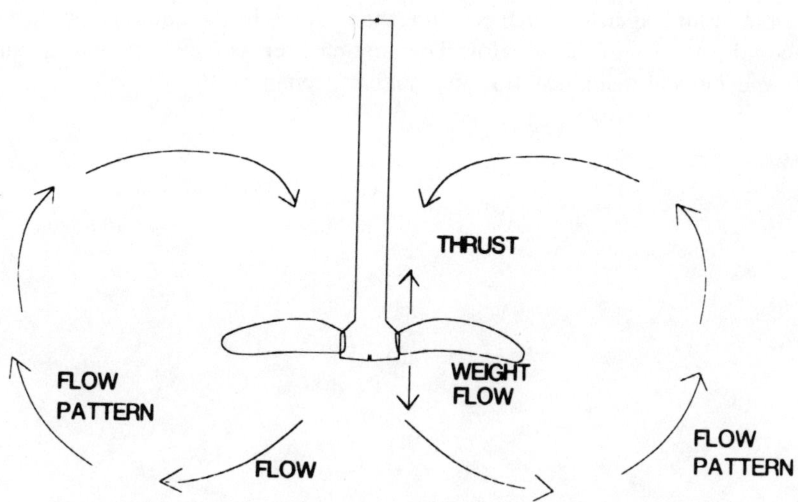

FIGURE 6-4. Thrust and weight forces with downpumping flow pattern.

greater than conventional axial or radial flow turbines and since the diameter of retrofit hydrofoils increase at constant powers draw, fluid forces will be significantly greater with retrofitted hydrofoils.

Fluid forces act perpendicular to the mixer shaft and create a bending moment when transmitted along the shaft. Bending moment is the product of fluid force multiplied by a distance along which the force is transmitted. Greater fluid forces will increase bending moments and therefore reactive forces acting on bearings, gears, mounting structures, and any mixer components along the shaft. As a result, a check must be made to see if the increased bending moment will exceed allowable limits of radial loads on bearings, gears, and mounting plates or flanges.

Shaft deflection is directly proportional to fluid force. Retrofitted hydrofoils will inherently cause higher shaft deflection due to greater fluid force. A calculation of shaft deflection at mechanical seals or stuffing boxes is necessary to be sure that it does not exceed allowable limits and cause failure of the seal or stuffing box. The mixer supplier should perform this check to be sure that the retrofit will not create excessive shaft deflection.

Shaft shear stress is a combination of torsional and bending moment stresses. Retrofitted hydrofoils will impose greater shaft stress due to higher bending moment. Shaft stress due to bending is directly proportional to bending moments. Shaft shear stress is the square root of the sum of the squares of bending and torsional stress. Therefore, a calculation must be made pertaining to shaft shear stress to ensure that allowable stress limits are not exceeded with the retrofit. Such an event could deform or break the mixer shaft.

Retrofit hydrofoils will improve process results or reduce power consumption, but at the expense of increased mechanical loads that could jeopardize the structural integrity of the agitator. The next chapter will show how to quantify these mechanical loads resulting in a sound mechanical design.

7

Mechanical Design of Agitators

Once an agitator has been designed from a process perspective, it must be designed mechanically. In other words, after the power has been calculated and the type and diameter of impeller and operating speed have been chosen to achieve the desired process result, the mixer shaft diameter and length and drive assembly must be selected to complete the mixer design. Impeller blade thickness must also be specified. Shaft design, drive train selection and impeller blade thickness specification incorporate the ''mechanical design'' of the mixer.

The mechanical design of an agitator cannot be qualified in whole until the magnitude and nature of the forces on the agitator are understood. These forces are the result of fluid dynamics generated by the interaction of the impeller with the internals of the vessel and the specific process involved. These produce both constant and variable stresses on the agitator assembly. Before a complete understanding of the interrelationship of the mechanical design and these forces can by completely evaluated, the following must be considered:

1. The magnitude and nature of the forces must be identified
2. The conceptual design of the agitator must take into account these forces
3. The engineering standards and manufacturing tolerances must be consistent with the design of the equipment

The fluid forces acting on an agitator can be grouped as follows. Those related to the fluid dynamics produced by the agitator and the interaction with the vessel internals. These forces are normally attributed to baffles, impeller location, and/or the energy density input of the mixer. Another group of fluid forces arises from the process. These would include forces from violent reactions, gas dispersion and/or reactions containing a high concentration of solids. Prediction of these forces is usually dependent upon literature searches, knowledge of the specific process involved, and data from the various mixer suppliers.

The nature of the forces imposed on agitators normally dictates the design required to offset and withstand these forces. For instance, mild agitation produces quantifiable fluid forces and is achieved by directly connecting the mixer shaft to a speed reducer (gearbox) that has been designed to withstand loads. If the dynamic loads are significantly high and/or are difficult to measure or quantify by prediction or calculation, a design predicated on isolation or dampening of these destructive forces from the drive assembly may be in order. Since the drive train (gearbox) often represents the greatest component cost of the mixer assembly, it is particularly important to protect the gearbox from these fluid forces.

Consideration must also be given to the engineering standards and manufacturing tolerances used and adhered to by specific agitator suppliers. Those practices must be in keeping with sound engineering principles and practices. They must be consistent with the assumptions used in the calculation and design of the equipment.

A. SHAFT DESIGN

Factors involved in agitator shaft selection and operation include data plus methods for determining allowable shaft lengths. These factors will be presented in this section.

For proper shaft design, an agitator shaft must satisfy the limitations imposed by:

1. Critical speed and vibration
2. Shaft stress
3. Shaft deflection

A shaft of specific length must have a minimum diameter to satisfy these criteria optimally with respect to shaft design.

B. CRITICAL SPEED AND SHAFT VIBRATIONS

The most important factor in determining allowable shaft length is usually critical speed. Critical speed of a rotating shaft is any speed at which the forces acting on the shaft give rise to large vibrations.

Critical speed is defined as that speed at which the rpm of the shaft is equal to its natural lateral vibration frequency measured in cycles per minute. When the operating speed of a rotating shaft approaches the natural frequency, it will vibrate with increasing amplitude, such that shaft failure may occur due to excessive bending stress or by exceeding the yield point of the metal such that permanent deformation or shaft breakage occurs. Critical speed is a mechanical consideration dependent solely on materials of construction, shaft diameter, shaft length, bearing spacing in the drive assembly and impeller weight.

In practical terms, the critical speed of a shaft defines operating speed ranges that are impossible for the agitator to sustain. Critical speeds are the natural harmonic frequencies of bending in the shaft. If the shaft were excited with an oscillating force with a frequency close to the shaft harmonic, the shaft will store most of the energy of the exciting force by increasing the amplitude of the vibration until the shaft bends or breaks.

It is extremely important to be able to predict the vibration performance of the mixer shaft in order to compensate for potential vibration problems before any equipment is actually installed. A vibration problem manifests itself in high maintenance costs, frequent downtime and premature replacement of shaft bearings, shaft seals, oil seals and gears in speed reducers, etc. There are several ways a shaft can vibrate, but bending vibration (termed the ''critical speed'') is the most destructive factor when it comes to shaft deformation or failure.

Three types of vibrations can occur in a mixer shaft and these are:

1. Axial vibrations
2. Torsional vibrations
3. Bending vibrations

Axial shaft vibrations are the imperceptible lengthening and shortening ''accordion effect'' of the shaft along its centerline. The axial natural frequencies are usually so high (>3000 Hertz) that in most practical installations, machine and/or fluid excitation forces rarely produce much vibration.

Torsional forces are more serious in that they cannot be perceived by the human senses and are difficult to measure. This twisting vibration of the shaft will usually manifest itself in the form of shaft failure or gearbox gear failure.

Bending vibrations are the most serious form of vibration; they are generally referred to as critical speed. If the shaft is excited by an oscillating force with a frequency which corresponds closely to one of the critical speeds, the agitator shaft will bend excessively and the support structure will be seen to vibrate. Usually because of nonsymmetrical support structure flexibility, the motion which occurs is a vibration in the plane on which the structure is most compliant.

In theory there are an infinite number of critical speeds or harmonic frequencies in any given shaft system. The lowest frequency at which a shaft can vibrate is called the first critical speed, the next lowest frequency is called the second critical speed, and so on. As a general rule the higher the natural frequency the less destructive energy that can be stored in the shaft associated with that frequency. In practical terms, critical speeds higher than the second are of no concern.

The models used in the calculation of critical speeds assume an infinitely rigid mounting structure. Compliant mounting structures, such as insufficiently rigid bridges or flexible fiberglass nozzles, will invalidate the infinitely rigid mounting assumption, resulting in an overestimation of the installed critical speed. There are other imperfections in the models, such as the assumption of perfectly stiff bearings, and omissions of gyroscopic thrust and tensile forces. It is important to note that the calculated critical speed is not the installed critical speed. The installed critical speed is always lower. This is why mixer operating speeds should not exceed 70% of the calculated first critical speed. In extreme cases where the mixer mounting is flexible, the supporting flange may be gusseted for added strength and rigidity or the tank top may be strengthened to compensate for excessive vibrations.

Shaft critical speed N_{cr} can be calculated as follows:

$$N_{cr} = K_1 (d/L)^2 \sqrt{\frac{E}{\left(\frac{L+a}{L}\right)\left(W + \frac{(K_2 \times W_e)}{L}\right)}} \qquad (1)$$

where:
N_{cr} = shaft critical speed, rpm
K_1, K_2 = constants
d = shaft diameter, inches
L = shaft length, inches
E = modulus of elasticity, $lb/inch^2$ (see Table 7-1)
W = shaft weight per unit length, lb/inch (see Tables 7-1 and 7-2)
W_e = equivalent impeller weight, lb (see Fig. 7-1)
a = bearing spacing in the gearbox, inches

From Eq. (1), careful consideration must be given to material of construction for shafts, the impeller equivalent weight, shaft diameter, shaft length and bearing spacing. Critical speed varies inversely as the square root of the ratio of impeller equivalent weights. Therefore a heavier shaft lowers the critical speed of a shaft and reduces the allowable shaft length for operation at a given speed.

Of greater importance is the effect of changes in incremental shaft length and shaft diameter on critical speed. A change in either factor changes the critical speed by the square of the respective ratio. Thus, while changes in impeller equivalent weight and material of construction (i.e., modulus of elasticity) do have an effect on critical speed, the magnitude is small. Changes in shaft length and/or diameter has a large effect on critical speed.

TABLE 7-1. Reference Solid Shaft Weight Data for Critical Speed Calculations.

Shaft Dia.	Carbon Steel	Stain-Steel 304/316	Monel Alloy 400	Nickel Pure	Inconel Alloy 600	Hastelloy-B	Hastelloy-C	Carpenter-20	Titanium Pure	Aluminum Pure
Density:	.284	.290	.319	.322	.304	.334	.323	.269	.163	.098
Modulus:	30	28	26	30	31	30.8	30.9	28	15	10
1″	.22	.23	.25	.25	.24	.26	.25	.23	.13	.08
1.25″	.35	.36	.39	.40	.37	.41	.40	.35	.20	.12
1.5″	.50	.51	.56	.57	.54	.59	.57	.51	.29	.17
2″	.89	.91	1.00	1.01	.96	1.05	1.01	.91	.51	.31
2.5″	1.39	1.42	1.57	1.58	1.49	1.64	1.59	1.42	.80	.48
3″	2.01	2.05	2.25	2.28	2.15	2.36	2.28	2.04	1.15	.69
3.5″	2.73	2.79	3.07	3.10	2.92	3.21	3.11	2.78	1.57	.94
4″	3.57	3.64	4.01	4.05	3.82	4.20	4.06	3.63	2.05	1.23
4.5″	4.52	4.61	5.07	5.12	4.83	5.31	5.14	4.60	2.59	1.56
5″	5.58	5.69	6.26	6.32	5.97	6.56	6.34	5.67	3.20	1.92
5.5″	6.75	6.89	7.58	7.65	7.22	7.94	7.67	6.87	3.87	2.33
6″	8.03	8.20	9.02	9.10	8.60	9.44	9.13	8.17	4.61	2.77
6.5″	9.42	9.62	10.59	10.68	10.09	11.08	10.72	9.59	5.41	3.25
7″	10.93	11.16	12.28	12.39	11.70	12.85	12.43	11.12	6.27	3.77
7.3″	11.52	11.77	12.94	13.06	12.33	13.55	13.11	11.73	6.61	3.98
7.5″	12.55	12.81	14.09	14.23	13.43	14.76	14.27	12.77	7.20	4.33
200 mm	13.83	14.12	15.53	15.68	14.60	16.26	15.73	14.07	7.94	4.77
220 mm	16.73	17.09	18.80	18.97	17.91	19.68	19.03	17.03	9.60	5.77
240 mm	19.91	20.33	22.37	22.56	21.32	23.42	22.65	20.26	11.43	6.87
260 mm	23.37	23.87	26.25	26.50	25.02	27.49	26.58	23.78	13.41	8.06

Density in lb/in.^{3}
Modulus in psi × 1000
Weights in lb/in.

TABLE 7-2. Reference Hollow Shaft Weight Data for Critical Speed Calculations.

Pipe, nominal in. and schedule	Equivalent Shaft Diam., in.	Wall Thickness, in.	O.D., in.	I.D., in.	Weight per foot, lb/ft
4SCH80	3.52	.337	4.500	3.826	14.98
5SCH80	4.23	.375	5.563	4.813	20.78
6SCH40	4.42	.280	6.625	6.065	18.97
6SCH80	4.99	.432	6.625	5.761	28.57
8SCH40	5.55	.322	8.625	7.981	28.55
8SCH60	5.94	.406	8.625	7.813	35.64
8SCH80	6.30	.500	8.625	7.625	43.39
10SCH40	6.73	.365	10.750	10.20	40.48
10SCH60	7.38	.500	10.750	9.750	54.74
12 STD	7.64	.375	12.750	12.000	49.56
10SCH80	7.74	.593	10.750	9.564	64.33
12SCH40	7.83	.406	12.750	11.938	53.53
14 STD	8.16	.375	14.000	13.250	54.57
12 XHVY	8.33	.500	12.750	11.750	65.42
14SCH40	8.55	.438	14.000	13.124	63.37
12SCH60	8.62	.562	12.750	11.626	73.16
14 XHVY	8.90	.500	14.000	13.000	72.09
16 STD	8.94	.375	16.000	15.250	62.58
12SCH80	9.12	.687	12.750	11.376	88.51
14SCH60	9.35	.593	14.000	12.814	84.91
18 STD	9.70	.375	18.000	17.25	70.59
16SCH40	9.77	.500	16.000	15.000	82.77
14SCH80	10.00	.750	14.000	12.500	106.10
20 STD	10.43	.375	20.000	19.250	78.60
16SCH60	10.59	.656	16.000	14.688	107.50
18 XHVY	10.60	.500	18.000	17.000	93.45
16SCH80	11.39	.843	16.000	14.314	136.50
20 XHVY	11.41	.500	20.000	19.000	104.10
24 STD	11.81	.375	24.000	23.250	94.62
24 XHVY	12.93	.500	24.000	23.000	125.50

The complexity of equations used to solve for critical speed increases rapidly as the number of shaft diameter changes or "steps" in the shaft increases. Most critical speed calculations involve some form of approximation to reduce the complexity.

Equation (2) below is another form of Eq. (1) and can be used to calculate the shaft critical speed for a single stepped shaft. It allows for solid or hollow shafts. All weights such as impellers, couplings, etc. are transferred to the end of the shaft by the equivalent weight equation (3). Equation (3) is applicable to either (a) an unstepped shaft different in diameter from the speed reducer shaft

Shafts: Basis of Design

Critical Speed Calculation

$$N_{C_1} = 146.4 \frac{da^2}{L} \sqrt{\frac{E}{\left[(L+a)L + \frac{(RS-1)c^3}{L}\right]\left[4.13 \frac{W_e}{L} + W_B\right]}}$$

N_{C_1} = First Shaft Critical Speed (RPM)
d_A = Upper shaft outside dia. (Inches)
L = Shaft Length (Bearing to ℄ Impeller) (Inches)
E = Modulus of Elasticity
a = Bearing Span (Inches)
R = I_A / I_B A is Upper Shaft Moment of Inertia, B is Lower Shaft Moment of Inertia
S = Weight Factor S = 1 for 316, 304 & C.S.
C = Lower Shaft Length ($C \geq .75L$)
W_e = Equivalent Weight (Pounds)
W_B = Pounds Per Inch of Lower Shaft

Equivalent Weight (We)
$W_e = W_1 + W_2 \left(\frac{L_2}{L_1}\right)^3 \ldots.$

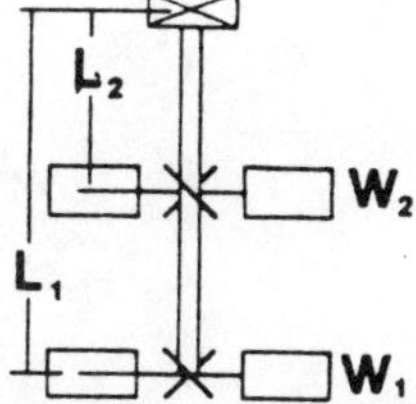

FIGURE 7-1. Equivalent weight concept for dual turbines. *(Eastern Mixers, Inc.)*

or (b) a single stepped shaft with the upper diameter equal to the speed reducer output shaft diameter and the lower shaft length is at least 75% of the overall shaft length from lower bearing to centerline of the lower impeller.

$$N_{cr1} = 146.4 \frac{da^2}{L} \sqrt{\frac{E}{\left[(L+a)L + \frac{(RS-1)c^3}{L}\right]\left[4.13 \frac{W_e}{L} + W_B\right]}} \qquad (2)$$

where: N_{cr1} = First shaft critical speed, rpm
d = Upper shaft outside diameter, inches
a = Bearing span in the gearbox, inches
L = Shaft length (lower bearing to impeller centerline), inches

E = modulus of elasticity, psi (See Table 7-1)
$R = I_A/I_B$, where I_A = Upper shaft moment of inertia, inch^4
I_B = Lower shaft moment of inertia, inch^4
$I \propto a^4$
S = Weight factor: $S = 1$ for 316/304 stainless steel and carbon steel
C = Lower shaft length ($C \geq .75L$)
W_e = Equivalent weight, lbs. (See Equation (3))
W_B = Weight per linear inch of lower shaft (lb per inch). (See Tables 7-1 and 7-2)

Equivalent Weight W_e can be expressed as

$$W_e = W_1 + W_2 \left(\frac{L_2}{L_1}\right)^3 + \cdots \tag{3}$$

(see Figure 7-1)

$$W_e = W_1 \times (L_1/L)^3 + W_2 \times (L_2/L)^3 \tag{4}$$

$$C_1 = 1{,}694{,}000 \times SD \times (L/(L + a))^{1/2} \tag{5}$$

$$C_2 = ((2a)/L)^3 + 33(L/a) + 4L \tag{6}$$

$$C_3 = (L/99) \times (a/(a + L))^2 \tag{7}$$

$$C_4 = (19W_e)/(SD^2 \times L) \tag{8}$$

$$C_5 = (L^2 \times (L + (C_2)(C_3) + C_4)^{1/2}) \tag{9}$$

$$N_{\text{cr1}} = C_1/C_5 \tag{10}$$

To insure safe operation, the actual operating speed should be less than 70% of the calculated critical speed (first critical), but outside the range of 45 to 55% of calculated critical speed (see Fig. 7-2). When a stabilizing ring or fins are incorporated on the impeller, the shaft may operate up to 80% of first critical speed. The additional weight of the fins or ring will reduce the critical speed, creating a situation where a loss of all or part of the stabilizing devices may result in catastrophic shaft failure.

Practical Critical Speed Design Methods

In practice one usually will calculate the first critical speed and the critical speed ratio (N_{cr}). The following guidelines have been established based on N_{cr}. An

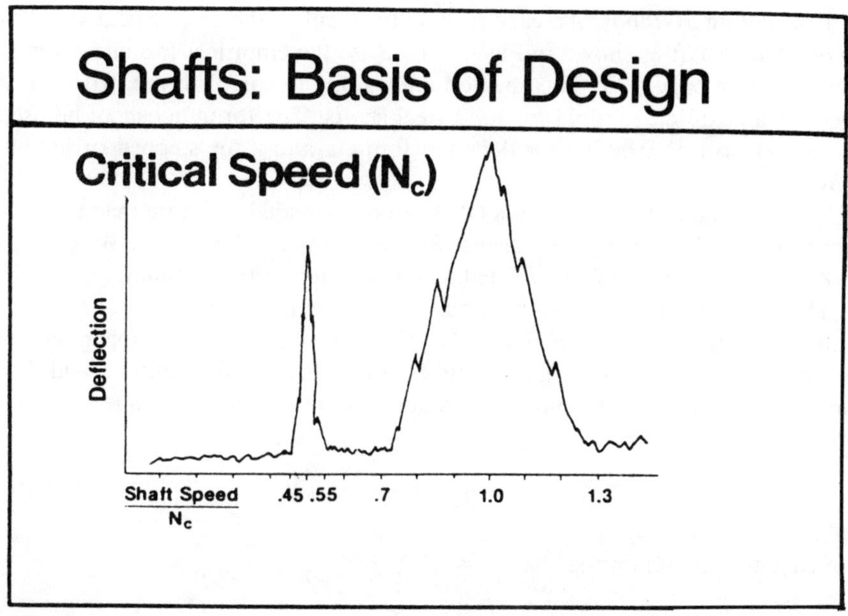

FIGURE 7-2. Ratio of operating to critical speed as a function of shaft deflection. *(Eastern Mixers, Inc.)*

overhung shaft is supported by two bearings in the gearbox housing and the end of the shaft is unsupported. A steady bearing shaft has the end of the shaft fitted into a tightly fitting bushing which acts like a bearing. In designing for shafts running in steady bearings be sure that the machine can operate in a stable overhung condition, just in case the steady bearing fails.

Overhung Shafts:

N_{cr} = 0–.5	No stabilizer required on the turbine.
N_{cr} = .5–.8	Stabilizer is added to increase dampening in the form of a ring or fins for axial flow impellers.
N_{cr} = .8–1.25	DANGER. Do not operate in this range
N_{cr} = 1.25–2.0	Stabilizer required.
N_{cr} = 2.0–Up	Calculate second critical speed.

Steady Bearing Shafts:

N_{cr} 0–.5	No stabilizer required, but check N_{cr} for overhung shaft and make sure that N_{cr} not in .8 to 1.25 range in case the steady bearing fails.
N_{cr} .5–Up	Refer to mixer supplier for design help.

Calculating Methods. An easy method to calculate the first critical speed of an overhung shaft as shown in Fig. 7-3 is to use the empirical formula given in Eqs. 4–10 on page 192. This method allows one to calculate the first critical speed in an explicit formula for solid steel shafts. The formula cannot be used for higher critical speeds, nor does the formula allow for stepped or hollow shafts.

The equivalent weight formula (W_e) can be extended for more than two turbines by the addition of more terms. For the simple turbine case, W_e reduces to W. Shaft couplings can be treated as turbines for better accuracy.

The following formula can be used to approximate the first critical speed of a steady bearing mixer (see Fig. 7-4). The model assumes the upper section between the bearings in the gearbox to be a cantilever end condition, and thus the bearing spacing is unimportant. Shaft couplings can be included and are considered like a turbine.

$$C_0 = 7{,}306{,}577(SD)/L^2 \tag{11}$$

For each turbine (i) on the shaft, $i = 1, 2, \cdots,$

$$A_i = 17{,}080{,}000(SD^4)(L^3) \tag{12}$$

$$B_i = W_i(L - L_i)(L_i^3)(4(L^2) - 5(L_i)(L) + L_i^2) \tag{13}$$

$$C_i = 187.7(A_i/B_i)^{1/2} \tag{14}$$

$$N_{cr1} = \frac{1}{\left((1/C_0^2) + \sum_{i=1}^{n} 1/C_i^2\right)^{1/2}} \tag{15}$$

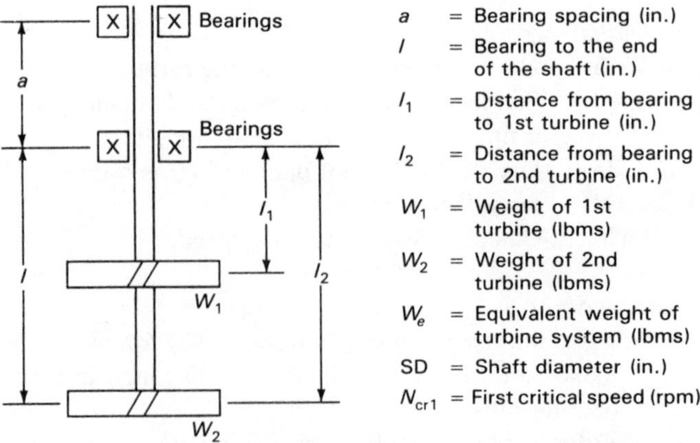

FIGURE 7-3. Typical overhung mixer shaft. *(Prochem Mixing Equipment, Inc.)*

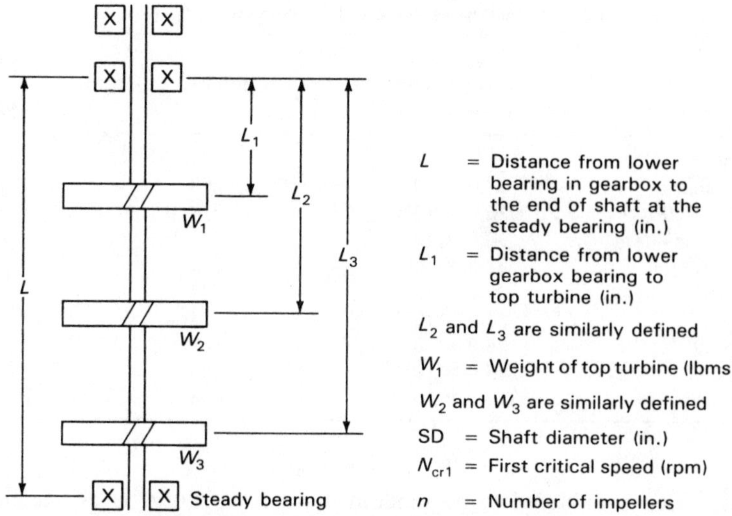

FIGURE 7-4. Typical steady bearing mixer shaft. *(Prochem Mixing Equipment, Inc.)*

The formula for N_{crl} is shown above for a 3 turbine system. The formula can be extended for a fourth turbine by the addition of C_4 calculated in the same manner as C_1 through C_3 and included as the other terms in the N_{crl} expression.

These two methods are valid only if the mixer shaft is constructed of carbon steel. Often a different metallurgy is requested which invariably has different properties than carbon steel. Properties such as density and modulus of elasticity are important in the critical speed calculation. The method for calculating critical speed for non carbon steel materials is as follows:

1. Determine the weight of the turbines in carbon steel using the same thicknesses as used with the new metallurgy. For example: Titanium turbine will be constructed to weight 35.0 lbms, therefore use the carbon steel equivalent weight in the calculations, 35 × (.284/.163) = 61 lbms (density ratio.) See Table 7-3
2. Calculate the first critical speed as illustrated previously using the carbon steel weights.
3. From Table 7-3, locate the critical speed multiplier for the metallurgy and multiply the critical speed in step 2 by the factor to obtain the true critical speed.

Excitation Forces

The major excitation forces on a mixer shaft all have frequencies which are some multiple of the operating speed of the shaft. The forces imparted to the

TABLE 7-3. Properties of Various Materials of Construction.

Metal or Alloy	Modulus of Elasticity, psi	Density, lbm/in.^{3}	Critical Speed Multiplier
Carbon steel	30.0×10^{-6}	.284	1.0000
304/316 stainless	28.0×10^{-6}	.290	.9560
Monel alloy 400	26.0×10^{-6}	.319	.8784
Pure nickel	30.0×10^{-6}	.322	.9391
Inconel alloy 600	31.0×10^{-6}	.304	.9825
Hastelloy-B	30.8×10^{-6}	.334	.9343
Hastelloy-C	30.9×10^{-6}	.323	.9516
Carpenter 20	28.0×10^{-6}	.289	.9577
Pure titanium	15.0×10^{-6}	.163	.9334
Pure aluminum	10.0×10^{-6}	.098	.9828

shaft are often considered to be random which, in theory, is all frequencies being represented to some extent. However, the random forces have the greatest magnitudes at frequencies near the multiples of the operating speed.

The most important excitation frequency is the operating speed itself because any mechanical imbalance in the shaft will result in a lateral force. The blade passing frequency is also extremely important. It is the number of impeller times the operating speed. For example, a 3-bladed turbine operating at 350 rpm was passing through drawoff and will strike the liquid surface 3 times per revolution, thus imparting an excitation force to the shaft with a frequency of 1050 rpm (17.5 Hz). The 1050 rpm was 90% of the calculated second critical speed and the shaft bent (remember the installed critical is always lower). It could be concluded that the blade passing frequency excited the shaft at its second installed critical speed. Similarly, side-entry mixers with 3-bladed turbines installed close to the tank bottom will excite the shaft at 3 times the operating speed. Mixers running at 33% of the first critical often have bearing failures. The number of blades times the operating speed is the excitation frequency to be concerned with.

C. SHAFT STRESS AND DEFLECTION

Shaft stress results from a combination of three loads. These are:

1. Torsional loads, those caused by the torque necessary to drive the impeller.
2. Axial loads, those due to the static weight of the shaft and impeller, and also the dynamic thrust of the impeller. Thrust loads are of importance and significance only with axial flow impellers.

3. Radial loads, those due to the hydraulic fluid forces resulting from the action of the fluid on the impeller, and from centrifugal or eccentric dynamic loads. Radial loads cause shaft deflection which in turn produces bending stresses.

Radial loads and bending stresses are important considerations in mixer shaft design. They are of significance when power input to the system is high, say greater than 10 Hp per 1,000 gallons, and the radial loads due to fluid forces are great.

Deflection is also a controlling factor, when it is necessary to have low shaft runout at the shaft seal (stuffing box or mechanical seal) or at the impeller.

Shaft deflections are the greatest during drawoff, when the impeller is rotating partly submerged, and the unbalanced loads are large. Deflections with the impeller operating fully submerged may also be significant, though less frequently observed.

Stabilizer rings and fins on blade tips are used to increase the allowable shaft length of a fixed diameter shaft operating at a fixed speed. These stabilizing elements do not alter the critical speed, but allow operating closer to the critical speed by dampening the effect of unbalanced hydraulic loads. Such loads are always present in operation and are larger through drawoff (i.e., when the liquid level passes the elevation of the impeller during filling and/or emptying). Stabilizers are most effective when operating totally in the liquid with the impeller blades partially submerged. Shaft whipping and vibration is minimized in this situation.

The magnitude of the shaft deflection depends on the ratio of the operating speed to the critical speed (N_{cr}), the amount of dampening in the system, and the magnitude of the periodic hydraulic forces. The hydraulic force calculation will yield the average magnitude of the force, but the deflection of the shaft cannot be calculated by static analysis alone. The actual deflection also depends on dampening and the critical speed ratio (N_{cr}).

Dynamic Deflection

The dynamic deflection of a shaft cannot be calculated by static deflection techniques alone. The actual deflection is a function of the critical speed ratio, dampening, and the magnitude of the forces on the shaft. Refer back to Fig. 7-2 for shaft deflection as a function of critical speed ratio. A shaft operating in air will have practically no dampening (except for the inherent internal dampening). A shaft immersed in a fluid will have some dampening. The dynamic deflection divided by the calculated static deflection is the magnification factor. The magnification factor is a strong function of the critical speed ratio, N_{cr} which is the ratio of the operating speed to the first critical speed. For a unit operating in air in the N_{cr} range of 0 − 2.0, the magnification factor can be expressed as:

$$\text{Magnification factor} = \frac{1}{1 - N_{cr}^2} \tag{16}$$

The formula goes to infinity, when the shaft operates at the first critical speed, indicating that the shaft will break. A unit operating in water will have some dampening and will deflect less as a result. In that case it is expected that the magnification factor will be at least 2 at $N_{cr} = 1$. A stabilizer will increase the dampening reducing the magnification factor. The magnification factor is less than 1 when a unit is operated above $N_{cr} = 1.5$. This is because the shape of the vibrating shaft is in the form of an S-curve, with the maximum deflections in the middle of the shaft, as opposed to having the maximum deflection at the shaft end when N_{cr} is less than 1.5.

Shaft deflection at the mixer mounting flange or at the shaft seal (packing gland or mechanical seal) may be estimated as follows:

$$\delta = \frac{\text{HF} \times \text{FF}}{6EI} \left[(2 \times (l \times a) + (3l \times \text{FF}) - \text{FF}^2)\right] \tag{17}$$

where: δ = shaft deflection, inches
HF = hydraulic fluid force at the impeller, including impeller weight, lb-force
FF = distance from lower bearing in speed reducer gearbox to mounting flange, inches
E = modulus of elasticity, psi
I = moment of inertia, inch^4, where

$$I \propto d^4 \tag{18}$$

where d is shaft diameter
l = shaft length from mounting flange to impeller centerline, inches
a = gearbox bearing spacing, inches.

Allowable shaft deflections caused by bending moments for stuffing boxes should not exceed .030 inch, whereas for mechanical seals, deflections should not exceed .010 inch. Consult the mixer manufacturer for maximum permitted deflections.

Shaft Stress

Shear stress is the result of forces acting on a shaft and is the resultant of torsional and bending stresses. If the stress in the shaft is too large the shaft will deform permanently resulting in a broken or bent shaft. Shear stress is the second major design constraint on the agitator shaft next to critical speed.

The lower shaft must be designed to withstand the stresses developed as it is rotating the turbine. The torsional and bending stresses are resolved into a single shear stress. As a result of the bending moment forces acting on the shaft, the shaft will bend slightly. The shaft may be designed for the jammed condition in which the customer starts up the agitator in a bed of settled solids. Shaft stresses under this situation are abnormally higher than under normal operating conditions.

The shaft on a belt-driven mixer is always machined to a smaller diameter to pass through the outboard bearing to the driven sheave. This stub is subjected almost entirely to torsional stress, which is greatest for slow speed belt driven mixers. If the shear stress is too high at the stub, it may twist off.

The shaft stress section will be divided into two distinct levels of understanding. The first will present the current design procedures used to predict the shear stress and the deflection in a shaft in a cook book method. For those interested, the hydraulic force and modern mechanical engineering methods of fatigue analysis will also be discussed. A procedure which combines the critical speed results with stress results will be introduced to upgrade our current knowledge.

Current Design Methods

The current design procedure is to calculate the shear stress and compare it to a present criterion. If the shear stress is larger than the criterion, the shaft diameter must be increased to reduce the shaft stress below it.

The lower shafts are designed with a shear stress which must not exceed 6000 psi. The torque is converted to a torsional stress and average bending moments used to calculate the average bending stresses, which are resolved into a single shear stress. The calculated shear stress on the lower shaft is not to exceed 6000 psi on any mixer. The sheave stub on slow-speed gearbox–belt combination drive units can be stressed to 7500 psi, based on torsional stress alone.

The shear stress is not constant for the entire length of the shaft and maximum allowable shear stress limits apply to the entire length of shaft. For overhung shafts, the maximum shear stress is usually at the inboard bearing, except when the shaft is stepped down to a smaller diameter. In the case of a stready bearing unit, the maximum shear stress is usually at the upper turbine. The calculated shear stress at either of these two points must not exceed the 6,000 psi criterion.

Calculating Torsional Stress

Given the nameplate Hp and the final operating speed, the torque delivered to the shaft can be calculated. In the case of multiple turbines, the torque transmitted by the shaft will decrease suddenly below a turbine. In the current design methods, the entire motor Hp is assumed to be transmitted to the turbine shaft even though the usual design practice is to have 80% of the motor Hp transmitted.

$$T_q = \frac{63025\ \text{Hp}}{N} \quad \text{(in.-lbf)} \tag{18}$$

$$SS_t = \frac{5.093(\text{OD})\,(T_q)}{(\text{OD}^4 - \text{ID}^4)} \quad \text{(psi)} \tag{19}$$

where: Hp = Nameplate Hp of Motor (Hp)
ID = Inside diameter of shaft (in.)
N = Speed of output shaft (rpm)
OD = Outside diameter of shaft (in.)
SS_t = Torsional stress in shaft (psi)
T_q = Torque transmitted in shafting (in.-lbf)

Example a. 10 Hp belt driven mixer operating at 78 rpm with 3″ solid upper shaft, 2.25″ sheave stub. Calculate the torsional stress at the sheave stub, which is essentially the shear stress because there is no bending moment.

$$T_q = 63025(10)/(78) = 8080 \text{ in.-lbf}$$

$$SS_t = \frac{5.093(2.25)\,(8080)}{(2.25^4 - 0)} = 3613 \text{ psi}$$

Since the stub shear stress is less than 7500 psi the unit is within design practice.

D. HYDRAULIC FORCES

An undesirable feature of a rotating impeller is the hydraulic side force which is generated. The side force affects the design of auxiliary equipment, such as the shaft, speed reducer, and seal, as well as the mixer support structure.

As long as a proper balance is struck between allowable shear stress and the hydraulic side force, a properly designed shaft will result. The use of lower shear stress to compensate for low hydraulic side force may cause underdesign in the other support structures.

In order to calculate the bending stresses on a shaft, the hydraulic side force on the shaft must be known. The force acting on the shaft perpendicular to it at the impeller is called the hydraulic side force. The flow around a completely submerged turbine in a tank is generally nonsteady because of the turbulence which is produced by the turbine itself. The hydraulic force acting on the shaft is due to unbalanced forces acting on the blades of the turbine resolved into a net hydraulic force acting on the shaft. The hydraulic force cannot be predicted from theory, so empirical approaches have been developed.

The procedure requires judgment on the part of the engineer to determine the correct regime multiplier to use. The regime multiplier is a strong function of the flow pattern both in the vessel and near the turbine. The regime multiplier has been broken down into two components; the turbine force coefficient and the regime correction factor. Multiply these two factors together to get the regime multiplier. These two factors are given in tables on the following pages.

The greater the number of blades, the lower the hydraulic force. Also an axial turbine produces about twice the thrust of a radial flow turbine and hence greater forces are produced at the blades. These effects are accounted for in the turbine force coefficient.

The regime correction factors are a little more difficult to apply, since a given installation may have more than one regime correction factor. If more than one regime correction factor applies, multiply them together and if the result is greater than 5, then use 5. The highest the regime correction factor can be is 5. Each turbine on the shaft is assigned a regime correction factor which is seldom the same value.

If the mixer must operate while the tank is being filled or emptied the surface waves are generating additional lateral forces on the turbine. In this case, the shaft forces also depend on the coverage of the turbine. The fluid coverage over the turbine is a good criterion to identify whether the surface will be in a state of "no swirl," "swirl," or "vortex," and the first group of regime correction factors account for the additional lateral forces. The second group accounts for the mounting orientation of the agitator. Note that a regime correction factor of 5 is always used in side entry mixer applications, overriding any other possible effects; hence the balance of the correction factors are applicable to top entry. The hydraulic force is a direct result of the turbulence in the vessel, and the mixing intensity is the measure of that turbulence. The third group reflects the fact that the higher the mixing intensity, the greater the hydraulic force. The addition of a small amount of gas can dramatically increase the hydraulic force, especially if the turbine is to drive the gas bubbles to the bottom of the tank. The final group of regime correction factors are recommended in those cases where gas is being sparged under the turbine:

$$x = (\text{Regime Correction Factor})\,(\text{Turbine Force Coff}) \tag{20}$$

$$\text{HF} = 3.48 \times 10^{-4}(NP_0)\,(N^{1.67})\,(D/12)^{3.53}(\text{SG})\,(x) \tag{21}$$

where: D = Turbine Diameter (in.)
HF = Hydraulic force on shaft (lbf)
N = rpm of output shaft (rpm)
NP_0 = Power number of the turbine
SG = Specific gravity of fluid
x = Regime multiplier.

Example b. Single hydrofoil impeller, 19″ diameter trifoil blades, 350 rpm, 0.3 NP_0, 15″ off bottom, in a 6′ dia. × 6′ straight side flat bottom vessel and mounted on an angle riser, mixing intensity is "mild." Note an angle rise is a mixer mounting adapter flange, allowing for angular shaft entry into the tank. On top-entry mixers, an angle riser eliminates the need for baffles.

Determine QV level.

Determine Regime Multiplier:

Turbine force coefficient	1.2
No Swirl with +2D coverage	1.0
Angle Mounting	2.0
"Mild" mixing intensity	1.5
No Gas in Vessel	1.0
Regime Correction Factor (1.0 × 2.0 × 1.5 × 1.0)	3.0
Regime Multiplier (3.0 × 1.2)	3.6

$$HF = 3.48 \times 10^{-4}(.3)(350^{1.67})(19/12)^{3.53}(1.2)(3.6) = 40.5 \text{ lbf}$$

As previously stated, hydraulic force is a fluid force acting perpendicular to the shaft axis and results from a net imbalance of forces on the turbine. In this section, the hydraulic force will be reviewed and a greater understanding of these forces will be reached.

If the flow through the turbine were perfectly symmetrical (and if the turbine were made perfectly) then there would be no fluid forces. No matter what tolerances were used in the manufacturing of the turbine, flow will still be nonuniform as a result of flow separation (or turbulence) over the blades. As a result, there will always be some hydraulic force. The greater the number of blades, the lower the probability of a force imbalance, and axial flow turbines create greater thrust forces (and hence greater imbalances). The differences among the turbines are reflected in the turbine force coefficient and power number of the turbine in calculating the hydraulic force. Similarly the force imbalance is increased when the turbine draws a vortex (as in low coverage through drawoff) or gas is sparged into the turbines discharge. Hydraulic force increases are taken into account using the regime correction factors.

In the previous discussion, the hydraulic force was considered to be a constant static value. The fact of the matter is that the magnitude of the hydraulic force is not a constant value but varies over time. Furthermore, the hydraulic force acts on the shaft with well defined frequencies. The knowledge of these hydraulic forces with regard to frequency and amplitude is imperative for the mechanical design of agitators to provide troublefree operation.

The only way to measure the magnitude of the resulting force is to measure the magnitude of the shaft deflection. Keep in mind that the magnitude of the deflection is also a strong function of the critical speed ratio (N_{cr}). The average deflection is used to determine the average hydraulic force by using Hooke's law and modeling the shaft as an elastic spring.

The actual deflection will appear to vary randomly about some mean value; the mean value is the hydraulic force. For example, in light duty service, variations from the mean will be in the order of $\pm 30\%$, and may be as high as $\pm 100\%$ in severe duty service. Hydraulic force measurements were made to develop an empirical model, since it is impossible to predict them from theory. Fig. 7-5 is a sample strip chart commonly used to record the fluctuating hydraulic force. Note that there is an average force (hydraulic force, HF) and a time varying component of the force.

The strip chart contains more information than a simple eyeballed average of the hydraulic force. The appearing random forces (or deflections) can be analyzed with the mathematical technique of Fourier transforms, which can convert any curve to series of sine waves, each with a different frequency and amplitude. The technique determines an amplitude for each frequency, which in this case is the amplitude of the oscillated force. The result of the analysis is a force spectrum diagram, which clearly indicates the various forces associated with each frequency.

Invariably, the predominant force has a frequency equal to the blade passing frequency (No. blades × rpm). Fig. 7-6 depicts the force spectrum for a single three-blade turbine operating at 120 rpm (2.0 Hz).

Note the peaks at 120 rpm and 360 rpm (3 × 120). These exciting frequencies must not correspond to any of the critical speeds.

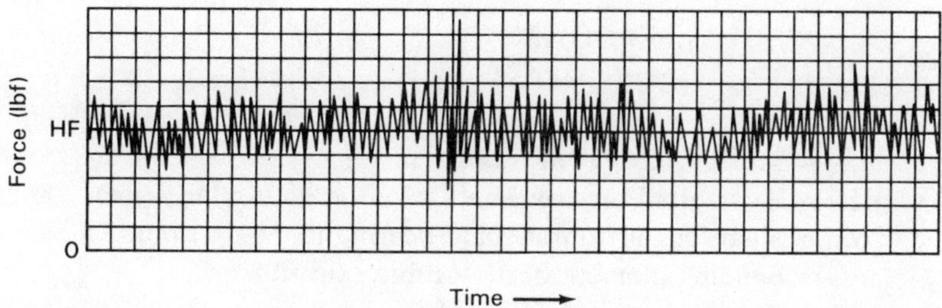

FIGURE 7-5. Hydraulic force variations over time. *(Prochem Mixing Equipment, Inc.)*

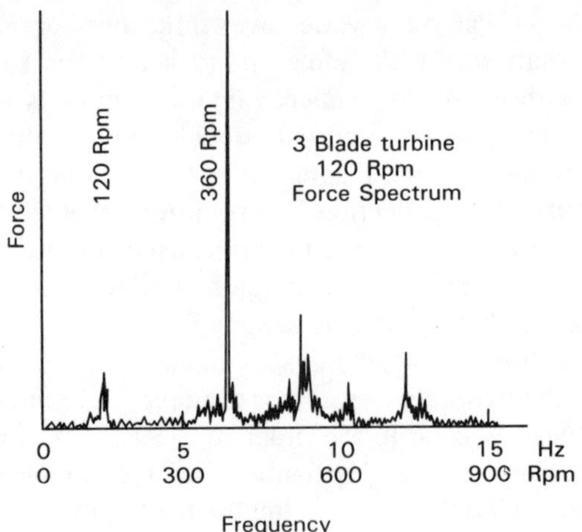

FIGURE 7-6. Force spectrum for a single three-bladed hydrofoil. *(Prochem Mixing Equipment, Inc.)*

Calculating Bending Stress—Overhung Shafts

The bending moment is greatest at the inboard bearing for overhung shafts. If the shaft diameter is constant throughout the entire length, the bending stress will also be the highest at that point. The shaft length from the inboard bearing is the shaft length from the mounting base to the turbine plus the mounting base to bearing distance.

The method below is used to calculate the bending moment at the inboard bearing, where the stress is usually the highest:

$$M_i = (\mathrm{HF}_i)(L_i) \quad \text{(in.-lbf)} \tag{22}$$

Total bending moment at inboard bearing:

$$M = M_1 + M_2 + M_3 + \cdots + M_n \quad \text{(in.-lbf)} \tag{23}$$

$$\mathrm{SS}_b = \frac{5.093(\mathrm{OD})(M)}{(\mathrm{OD}^4 - \mathrm{ID}^4)} \quad \text{(psi)} \tag{24}$$

where: ID = Inside Diameter of Shaft (in.)
HF_i = Hydraulic Force on Shaft from the ith turbine (lbf)
L_i = Shaft length from inboard bearing to the ith turbine (in.)
M_i = Bending moment of ith turbine (in.-lbf)
M = Bending moment (in.-lbf)
OD = Outside diameter of shaft (in.)
SS_b = Bending stress (psi)

Example c. Three identical 12″ diameter radial flow turbines with 6 blades and a center disk are mounted on a single 2.5″ diameter solid shaft at 50″, 75″, and 100″ from the inboard bearing. The particular turbine has power number of 5.3, and is to operate at 265 rpm with a 5 Hp motor. The ungassed fermenter medium has an SG = 1.05. Mixing intensity will be ''medium.'' The turbines will be able to disperse the gas that is used in the process.

Calculate the hydraulic force, then bending moment and bending stress:

$$\text{Regime Multiplier} = 2.5$$

$$\text{Hp} = (3.48 \times 10^{-4}(5.3)(265^{1.67})\,(12/12)^{3.53}(1.05)\,(2.5)$$

$$= 53.9 \text{ lbf}$$

$$M = (53.9)\,(50) + (53.9)\,(75) + (53.9)\,(100)$$

$$= 12,\ 127.5 \text{ in.-lbf}$$

$$SS_b = \frac{5.093(2.5)\,(12127.5)}{(2.5^4 - 0)} = 3{,}953 \text{ psi}$$

The torque at the inboard:

$$T_q = 1189 \text{ in.-lbf}$$

The torsional stress at the inboard:

$$SS_t = 387.6 \text{ psi}$$

The shear stress (SS) at the inboard bearing is the combination of torsional stress and bending stresses:

$$SS = (SS_t^2 + SS_b^2)^{1/2} = (3953^2 + 387.6^2)^{1/2} = 3972 \text{ psi} \qquad (25)$$

Since the shear stress is less than 6000 psi, the shaft is within design limits.

Example d . Could a 2″ diameter shafting be used for Example c)?

Changing the shaft diameter does not change the loads on the shaft, only the shear stress. Note that the shear stress for solid shafting is inversely proportional to the shaft diameter to the 3rd power for solid shafts. Using a 2.5″ diameter shaft, the shear stress calculated was 3972 psi.

$$SS = 3972(2.5/2)^3 = 7758 \text{ psi}$$

The 2″ shaft is unacceptable because it results in a shear stress greater than the 6000 psi criterion.

Bending Moment with Stepped Down Shafts—Overhung Shafts

If the shaft is stepped down, the bending moment should be calculated at the step as well. The turbines above the step on the overhung shaft do not contribute any bending moment to the stepped portion of the shaft below (nor do the turbines above contribute any torsion stress to the step). The bending stress is calculated based on the smaller (stepdown) shaft diameter. The bending stress at the step is calculated using the formula:

$$M_i = (\mathrm{HF}_i)(L_i - B) \quad \text{(in.lbf)} \tag{26}$$

$$M = M_i + M_{i+1} + M_{i+2} + \cdots + M_n \quad \text{(in.-lbf)} \tag{27}$$

where: B = Distance from step to inboard bearing, in.
i = First turbine below the step

Example e. Referring to Example C, if the shaft were stepped down from 2.5″ to 2″ at 60″ from the inboard bearing, what is the bending moment and shear stress at the step?

Two of the three turbines are below the step. They will contribute to the bending stress. Assuming that each turbine draws equal Hp, the torque at the step will be 2/3 that at the inboard bearing.

Calculating the moment at the step:

$$M = (53.9)(75 - 60) + (53.9)(100 - 60) = 2{,}964.5 \text{ in.-lbf}$$

$$T_q = (2/3)(1189) = 793 \text{ in.-lbf}$$

$$SS = \frac{5.093(\mathrm{OD})(M^2 + T_q^2)^{1/2}}{(\mathrm{OD}^4 - \mathrm{ID}^4)} \text{ (psi)} \tag{28}$$

$$\mathrm{SS} = 5.093\,\frac{(2)(2964.5^2 + 793^2)^{1/2}}{(2^4 - 0)} = 1954 \text{ psi}$$

The shaft stress does not exceed the maximum design constraint of 6000 psi total shear stress. (The last formula above is the combined stress formula.)

Calculating Bending Stress—Steady Bearing Shafts

The bending moment is usually the greatest at the upper turbine for steady bearing shafts. The bending moments are calculated at each turbine for the configuration shown in Fig. 7-7.

Calculate reaction force at steady bearing:

$$R_i = (\mathrm{HF}_i)\left(\frac{L_i}{L} - \frac{(2(L)(L_i) - L_i^2)(L - L_i)}{2(L^2)(L + a)}\right) \text{ (lbf)} \tag{29}$$

$$R_c = R_1 + R_2 + R_3 \text{ (lbf)} \tag{30}$$

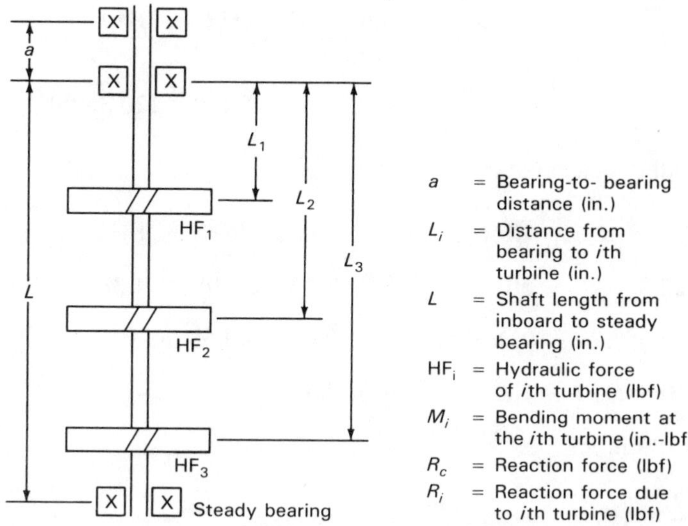

FIGURE 7-7. Typical steady bearing shaft with multiple turbines. *(Prochem Mixing Equipment, Inc.)*

Calculate the moment at each Turbine:

$$M_1 = R_c(L - L_1) - \mathrm{HF}_1(L_3 + L_2 - 2(L_1)) \quad \text{(in.-lbf)} \tag{31}$$

$$M_2 = R_c(L - L_2) - \mathrm{HF}_2(L_3 - L_2) \quad \text{(in.-lbf)} \tag{32}$$

$$M_3 = R_c(L - L_3) \quad \text{(in.-lbf)} \tag{33}$$

Calculate bending stress with same formula used in the overhung shaft case.

Example f). Referring to Example c, If a steady bearing were added to shaft arrangement at 120″ from the inboard bearing, what shaft diameter is required to keep the shear stress under 6000 psi.? (The bearing spacing is 15″)

$$R_i = (\mathrm{HF}_i)\left(\frac{L_i}{L} - \frac{(2(L)(L_i) - L_i^2)(L - L_i)}{2(L^2)(L + a)}\right) \text{ lbf}$$

$$R_1 = (53.9)\left(\frac{50}{120} - \frac{(2(120)(50) - 50^2)(120 - 50)}{2(120^2)(120 + 15)}\right) = 31.68 \text{ lbf}$$

$$R_2 = (53.9)\left(\frac{75}{120} - \frac{(2(120)(75) - 75^2)(120 - 75)}{2(120^2)(120 + 75)}\right) = 39.87 \text{ lbf}$$

$$R_3 = (53.9)\left(\frac{100}{120} - \frac{(2(120)(100) - (100^2)(120 - 100)}{2(120)^2(120 + 15)}\right) = 48.80 \text{ lbf}$$

$$R_c = R_1 + R_2 + R_3 = 31.68 + 39.87 + 48.80 = 120.35 \text{ lbf}$$

Calculate the moment at each turbine:

$$M_i = R_c(L - L_i) - \text{HF}_i(L_n + L_{n-1} + L_{n-2} + \cdots + L_{i+1} - (n - i)L_i)$$

$$M_1 = 120.35(120 - 50) - 53.9(100 - 75 - 2(50)) = 4{,}382 \text{ in.-lbf}$$

$$M_2 = 120.35(120 - 75) - 53.9(75 - 50) = 4{,}068 \text{ in.-lbf}$$

$$M_3 = 120.35(120 - 100) = 2{,}407 \text{ in.-lbf}$$

$$T_{q1} = 1{,}189 \text{ in-lbf}$$

$$T_{q2} = \left(\frac{2}{3}\right) 1189 = 793 \text{ in lbf}$$

$$T_{q3} = \left(\frac{1}{3}\right) 1189 = 396 \text{ in.-lbf}$$

The maximum torque and maximum bending moment is at the upper turbine, as initially guessed. Evaluating the shaft diameter for 6000 psi at the most highly stressed section of the shaft:

$$6000 \text{ psi} = \frac{5.093(M^2 + T_q^2)^{1/2}}{\text{OD}^3}$$

$$\text{OD} = (5.093(4382^2 + 1189^2)^{1/2}/(6000))^{1/3} = 1.568'' \text{ shafting}$$

Use 2″ shafting for steady bearing fermenter.

Calculating Shear Stress

The shear stress must be kept under 6000 psi for the lower shaft combining the bending stress and the torsional stress. The bending stress and the torsional stress are resolved into a single shear stress value by following formula:

$$\text{SS} = (\text{SS}_b^2 + \text{SS}_t^2)^{1/2} \text{ (psi)}$$

Several of the preceding examples make use of this method of combination of the two independent stresses to arrive at a single stress value. This concludes the cookbook section of stress analysis for shafts. On larger mixers where the time is warranted, more detailed approaches are being used. Tables 7-4 and 7-5 contain most of the data required to do a complete stress calculation of the shaft.

Stress on Jammed Shafts

In some applications, the turbine shaft may be jammed due to an abnormal situation, such as a solid suspension application with a mining slurry. The mill has a long power outage and the solids settle, covering the turbine. Later an operator tries to start the agitator and the agitator shaft fails.

Most motors have a starting torque which is 2 to 2.5 times the normal operating torque. The torque not coverted to horsepower causing flow is used to accelerate the rotation of the shaft. During startup and in the jammed condition, the torsional stress will be 2 to 2.5 times greater than the normal operating torsional stress.

In our current design methods, the sheave stub end can be stressed to 7500 psi based on the torsional stress under normal conditions. During startup or in the jammed condition, the maximum stress on the stub will be 15,000–18,750 psi. This should not be a problem since the yield strength of steel is 30,000 psi depending on the manufacturing methods used in making the steel. Thus the sheave stub end will never fail if the cookbook criteria is used.

The torque produced at the motor is transmitted to the jammed turbine where the large forces on the turbine blades counteract the torque supplied. If the solid

TABLE 7-4. Turbine Force Coefficients.

Turbine Style	Force Coefficient
Axial Flow	
Wide-blade hydrofoil with 5 blades	1.0
Wide-blade hydrofoil with 3 blades	1.2
3-blade propeller	1.2
6-blade propeller	1.0
45° Axial flow turbine with 4 blades	1.0
Radial Flow	
Flat-blade radial turbine with disk and 6 blades	0.5
Curved-blade radial turbine with disk and 6 blades	0.5
Bar turbine	0.6
Lifter turbine	0.6
Paddle 4 blades	0.9
Swirl backswept turbine	0.9
2 Bladed 45° Axial Flow Turbine and 2-Bladed Paddles	
1 Turbine on shaft	3.6
2 or 3 Turbines on shaft	1.8
4 or more turbines	0.9

TABLE 7-5. Regime Correction Factors.

Fluid Regime Description or Geometry in Tank		Regime Correction Factor
No Swirl	Well baffled and >2 turbine diameter coverage	1.0
Swirl	<1 Turbine diameter coverage and baffled	2.0
	Unbaffled no vortex	2.5
	Drawoff and surge tank	2.0
	Mixing intensity "medium" baffled	1.5
Vortex	<.5 Turbine diameter coverage	5.0
Off-center mounting with and without baffles		1.4
Angle mounting		2.0
Side entry units		5.0
Continuous flow and short residence times		1.5
Mixing intensity		
Mild		1.5
Medium		2.0
Vigorous		3.0
Gassed—fermentation		5.0
Gassed—flooded		2.0
Gassed at F = .005 ft/s and mixing intensity "mild"		4.0

bed is very firm, the motor will simply overload. The solid cake often breaks above the turbine, creating an opening upwards. The turbine rolls up through the opening on its blades, creating bending forces of the shaft. The distribution of forces on a jammed turbine are not known, but we will assume that all the jamming force acts at .75 of the turbine radius. From the jamming force, one can calculate the resulting bending moment.

$$F_{jam} = \frac{T_{jam}}{.375D} \quad \text{(lbf)} \tag{34}$$

$$M_{jam} = (F_{jam})\,(L) \quad \text{(in-lbf)} \tag{35}$$

where:
D = Turbine diameter (in.)
F_{jam} = Jamming force (lbf)
L = Shaft length from bearing to the turbine (in.)
M_{jam} = Bending moment in jammed condition (in.-lbf)
T_{qjam} = Jamming torque (2.5 times normal) (in.-lbf)

Example g. A 10 Hp drive is used on an agitator designed with a 54″ wide bladed hydrofoil with 5 blades with 3/16″ rubber covering, to operate at 52.1

rpm, with a 4″ diameter × 188″ long shaft. The specific gravity is 1.25 and the power number of the turbine is 1.10. (Data: 8.6″ from inboard bearing to mounting base, the regime multiplier is 1.0)

Turbine diameter is actually 55.125″ because triple thick rubber is applied to the tips.

The shaft length from inboard bearing to centerline of the turbine is 188″ + 8.6″ = 196.6″.

$$T_q = (63025)\ \mathrm{Hp}/N = 63025(10)/52.1 = 12097 \quad (\text{in.-lbf})$$

$$T_{q\mathrm{jam}} = (2.5)\ (T_q) = (2.5)\ (12097) = 30242 \quad (\text{in.-lbf})$$

$$F_{\mathrm{jam}} = \frac{T_{q\mathrm{jam}}}{.375D} = \frac{30242}{.375(55.125)} = 1463 \quad (\text{lbf})$$

$$M_{\mathrm{jam}} = F_{\mathrm{jam}})\ (L) = (1463)\ (196.6) = 287{,}626 \quad (\text{in.-lbf})$$

The shear stress in the jammed condition is calculated using the moment and torque in the jammed condition.

$$\mathrm{SS}_{\mathrm{jam}} = \frac{5.093\ (M_{\mathrm{jam}})^2 + (T_{q\mathrm{jam}})^2)^{1/2}}{\mathrm{OD}^3}\ (\text{psi}) \qquad (36)$$

$$= \frac{5.093(287{,}626^2 + 30242^2)^{1/2}}{4^3} = 23025\ (\text{psi})$$

This example shows that in the jammed condition, the shear stress can be as high as 23,000 psi, well above the 6,000 psi criterion. Since the jamming stress is less than the yield stress for steel, the shaft should not fail before the motor burns out. Note that the normal operating shear stress for this unit is expected to be only 1538 psi. Had the unit been designed with a 3″ diameter shaft, the normal operating shear stress would be only 3,645 psi within the design requirements. In the jammed condition, a shear stress of 54,554 psi will be developed which is greater than the yield stress and the 3″ diameter shaft would have bent permanently.

Stress and Critical Speed

The shaft stress calculations previously presented do not depend on the critical speed of the agitator. However, it is a fact that the critical speed ratio of an agitator shaft significantly affects the stress in the shaft. In this section, the effects of critical speed and shaft stress will be discussed.

The conclusion one can draw from the following is that the blade passing frequency (operating rpm × number of blades) should not match any critical speed of the shaft. For side entry, avoid the blade passing to the first critical, and with top entry avoid the blade passing to the second critical as well.

The calculated hydraulic force will be amplified by the critical speed ratio. The result is that the shaft will deflect more than calculated from statics alone. The magnification factor is also a function of the amount of dampening in the system. (The shock absorbers in a car are a dampening device.)

An agitator operating in air has practically no dampening. Although the hydraulic forces will be essentially zero, there will still be a small force due to the mechanical imbalance. Even this small force can destroy the shaft if the critical speed ratio is 1.0 because the magnification or amplification factor becomes extremely large. The amplification factor for no dampening is given as:

$$\text{Magnification factor} = \frac{1}{1 - N_{cr}^2}$$

The shaft fails because the force is magnified with each rotation of the shaft. Only after a long time will the force be magnified to its maximum value (as given by the factor). If the shaft was to operate above critical, then it will remain in the critical region only for a short period (.1 sec) and the force will not be magnified much in that short time. The critical speed will not break the shaft. The stresses developed as a result of the magnification of the forces exciting the shaft will break the shaft.

Any force which has a frequency equal to one of the critical speeds of the shaft will be amplified by the magnification factor given above. In the case of an unbalanced turbine, the exciting frequency is the operating speed of the shaft. In the case of hydraulic forces, the predominant frequency is the number of blades times the operating speed, called the blade passing frequency. The hydraulic force is larger by the magnification factor when the blade passing frequency and the closest critical speed are used in the formula. One may be surprised how seemingly fine operating mixers are overstressed. The shaft may fail, or the bearings will tend to wear out. The next three examples are case histories.

Example h. The following side entry unit has operated for 11 years with no bearing replacement and no operational problems. The unit used a 3 bladed propeller.

Critical speed calculated as	1066 rpm
Operating speed	230 rpm
Blade passing frequency	689 rpm
Shear stress	2500 psi
Magnification factor	$\dfrac{1}{1 - (689/1066)^2} = 1.72$

$$\text{Shear stress } (2500)(1.72) = 4300 \text{ psi}$$

The shear stress in the shaft is only 4300 psi.

Example i. The following side entry unit required 10 bearings, 3 bearing housings, and 1 shaft over 3 years of operation. A 3-bladed propeller was used.

Critical speed calculated as	1221 rpm
Operating speed	425 rpm
Blade passing frequency	1275 rpm
Stress	2230 psi

$$\text{Magnification factor} = \frac{1}{1 - (1275/1221)^2} = 11.0$$

$$\text{Shear stress } (2230)(11.0) = 24{,}530 \text{ psi}$$

The relative stress in the shaft is very high at 24,530 psi because the hydraulic force has a frequency close to the critical speed of the shaft.

Example j. The following is typical of documented cases in which the shaft broke during drawoff. The belt driven agitator mounted on an angle riser and a 3-blade turbine was used.

1st critical speed	101 rpm
2nd critical speed	730 rpm
Operating speed	229 rpm
Blade passing frequency	687 rpm
Stress at drawoff	5541 psi

$$\text{Magnification factor} = \frac{1}{1 - (687/730)^2} = 8.74$$

$$\text{Shear stress } (5541)(8.74) = 48{,}430 \text{ psi}$$

The shear stress in the shaft is very high at 48,430 psi, above or close to the yield stress of the material.

E. MIXER DRIVE DESIGNS

Gear Box Speed Reducer

A properly designed mixer gear drive is more than just a speed reducer, even though the speed reducer is a very important part of the drive assembly and usually is the most expensive component of the mixer. Speed reducers are typically 60% or so of the overall mixer purchase price.

Before a gearbox drive can be designed, the loads that the components (i.e.: bearings, gears, etc.) will be subjected to must be quantified. These include bending moment due to fluid force and torque due to rotational resistance. Bending moments are fluctuating loads of various amplitudes dependent upon vessel geometry and process conditions (ie: flow regimes, gassed or ungassed conditions, etc.). If the process requires mild agitation, for example, these forces are clearly defined if the mixer supplier has the proper test facilities and instrumentation to measure forces and bending moments on shafting. Strain gages and oscilloscopes can help quantify loads and bending moments. Torsional load is a straightforward calculation based on measured power draw (or motor nameplate power for a worse case scenario) and operating shaft speed.

Figure 7-8 depicts the bending moment seen by the mixer shaft, and thus the gearbox drive, under normal operation. The amplitude of these bending moments become questionable as the power per unit volume increases, and as the potential for a process upset increases. Consider a gas–liquid application where the amplitude of these fluctuations increases because the incoming gas produces additional vibrational and buoyant forces on the shaft and drive system.

A review of agitator gearbox drives commercially available for mixer applications is required. One type of gearbox drive has the agitator shaft directly coupled to the reducer bearings. This configuration is depicted in Fig. 7-9.

With this arrangement and drive design, the bending moments seen by the mixer shaft are also directly transmitted to the reducer shaft and bearings. The bearings are subjected to exceptionally high loads because of the relatively small bearing spacing, long overhung shaft, and large fluid forces as shown previously in Fig. 7-3.

This type of design can cause gear problems associated with high fluid forces and shaft deflection between the bearings in the gear reducer. Shaft deflection in the span between the bearings can cause the low speed gears to deflect, as illustrated in Fig. 7-10.

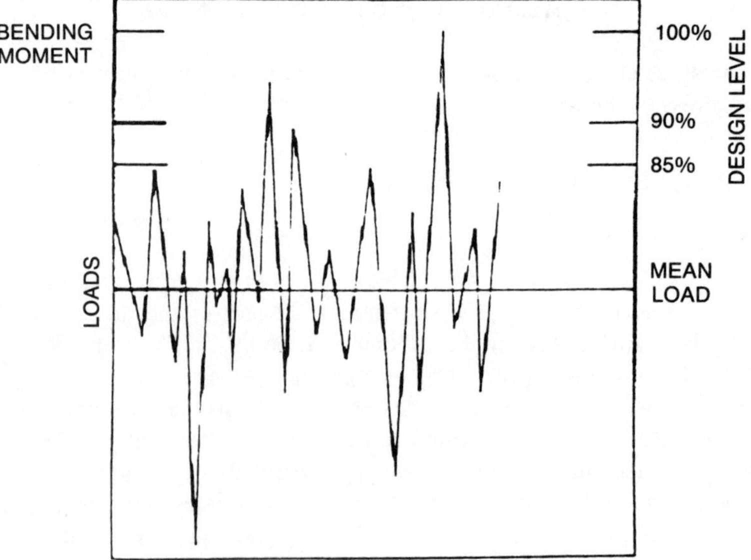

FIGURE 7-8. Bending moments transmitted to mixer drives. *(Mixing Equipment, Inc.)*

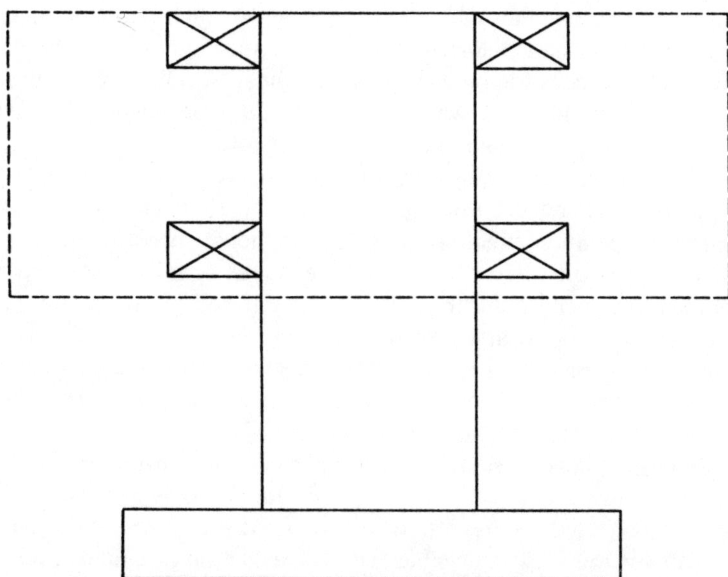

FIGURE 7-9. Shaft directly coupled to gear box bearings. *(Mixing Equipment, Inc.)*

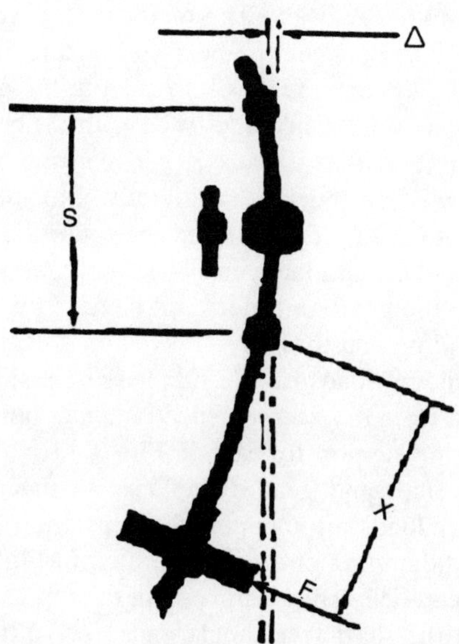

FIGURE 7-10. Deflection in directly coupled shaft. *(Mixing Equipment, Inc.)*

Excessive gear wear due to improper meshing of the teeth can occur. Increasing the reducer shaft diameter can decrease misalignment in the gearing due to shaft deflection, but in no way does it eliminate it. Directly coupled shaft assemblies are normally used when fluid forces, F, are quantifiable and the bending moments are well within allowable limits of the gearbox.

This type of design also transmits all of the torsional shock loads directly into the gearbox. Therefore, this type of drive and shaft arrangement should only be used in operations that do not include torsional shock loading.

An alternative to the directly coupled drive is a design that isolates bending moments from the speed reducer. This is accomplished by using either an independent bearing support and flexible coupling between the mixer shaft and reducer shaft, or by incorporating a belt drive between the mixer shaft and gear reducer. In the latter assembly the belt drive in essence acts like a flexible coupling. These designs are shown in Fig. 7-11.

The independent bearing support and flexible coupling design in Fig. 7-11(a) is commonly known as the "hollow quill" design. This is because the speed reducer portion of the drive has a hollow low speed shaft like a quill. Note the speed reducer section view shows the last gear reduction of a double reduction reducer. The low speed shaft (and thus low speed gear) is supported by its own set of twin bearings. The mixer shaft fits through the quill or hollow shaft and is supported by its own double set of bearings. The mixer shaft is connected to the speed reducer shaft by means of a flexible coupling as shown.

The combination belt drive/speed reducer assembly in Fig. 7-11(b) also isolates torsional loads from the gearbox. The mixer shaft is supported by its bearing housing. The belt drive between the bearing housing and gearbox acts like a flexible coupling and dampens the shock load from the gearbox. Load isolation from the gearbox can only be accomplished with the belt drive between the gear box and mixer shaft. If the gearbox is between the belt drive and mixer shaft the shock loads will be transmitted directly into the gearbox. The belt drive will accomplish a speed reduction and therefore the gearbox per se can usually be a relatively small (and inexpensive), single reduction unit. Since the loads are isolated from the gearbox it does not have to be designed to see high shock loads due to bending and torque.

In both deisgns, bending loads on the mixer shaft resulting from hydraulic fluid forces are isolated from the gearbox speed reducer portion of the drive due to (1) the separate bearing support for each in Fig. 7-11(a) and (2) the belt drive in between the mixer shaft and gearbox reducer in Fig. 7-11(b). Also peak torsional or horsepower loads are dampened out by the flexibility in each design. The isolation and damping effect is illustrated in Fig. 7-12.

Flexible couplings and belt drives dampen out the top 25–40% of peak shock loads. For example, if the shaft sees shock loads of two times normal running torque, then the flexible aspect of the coupling or belt drive will only transmit

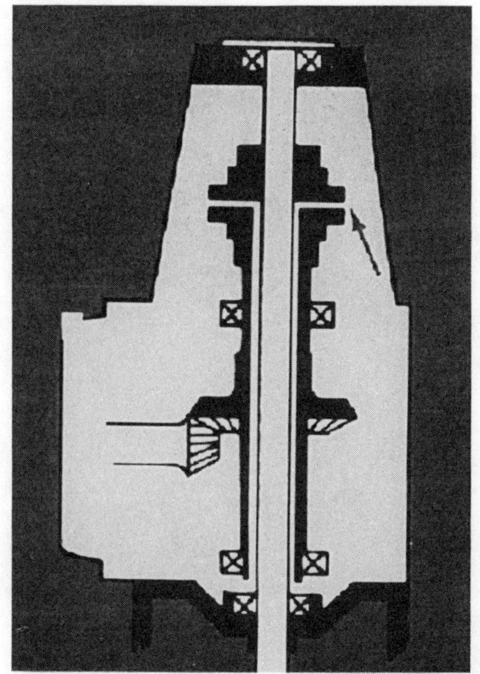

FIGURE 7-11(a). Hollow quill design for flexibly coupled shaft arrangements. *(Mixing Equipment, Inc.)*

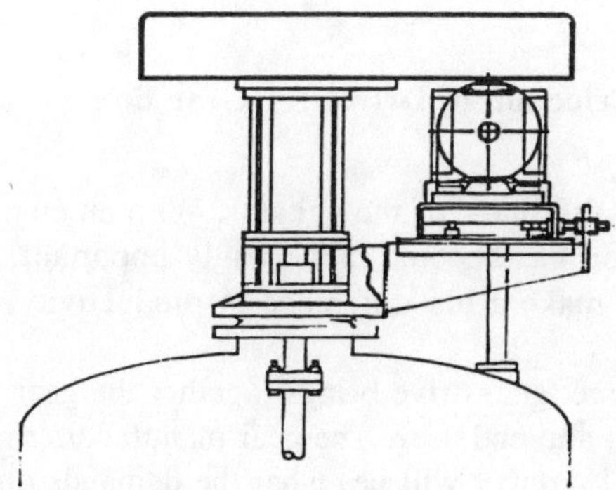

FIGURE 7-11(b). Belt drive/gearbox design for flexibly coupled shaft arrangement. *(Prochem Mixing Equipment, Inc.)*

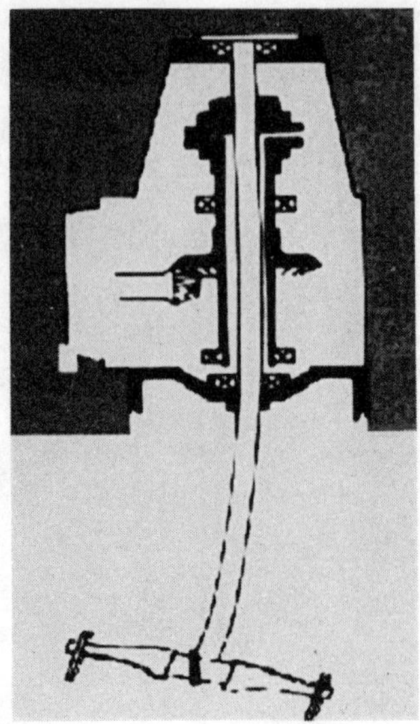

FIGURE 7-12. Flexible coupling drive design and load dampening and isolation effect. *(Mixing Equipment, Inc.)*

loads below 1.25 to 1.40 times normal running torque. This is illustrated in Fig. 7-13.

Both graphs show output torque variations with the same input load variation. Since the speed reducers are torsionally isolated and see much lower loads, the gearbox inherently has a greater service factor and longer life expectancy. A reducer having an AGMA service factor of 1.5 would have a corresponding service factor of 1.875 to 2.1 based on loads that were dampened out by a flexible coupling or belt drive.

Specification and Selection of Drives for Gear Box Speed Reducers

A gear drive is the "lifeline" of the agitator. With an emphasis on efficiency, proper drive selection has become increasingly important. The correct choice (or wrong one) can make a big difference in productivity, operating cost and energy savings.

Selecting the correct gear drive brings together the gear manufacturer, mixing system designer, and end user. The gear manufacturer must know what the end use of the speed reducer will be, what the demands on it will be, and the nature of the equipment the gearbox will be driving.

User and designer must be familiar with the variables that affect performance and service. For example, a mixing application that places a torque load on the

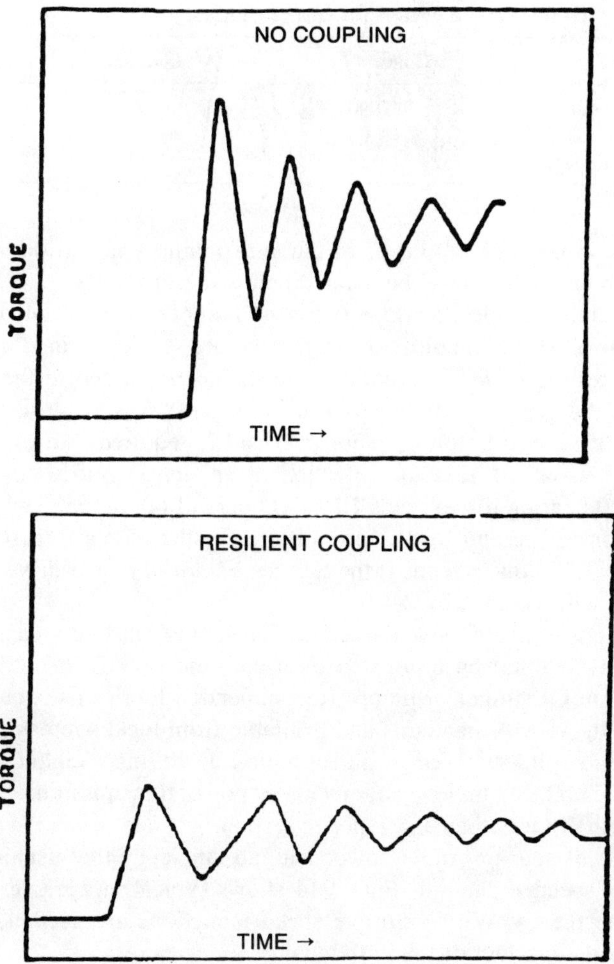

FIGURE 7-13. Torque dampening effect of flexible coupling or belt drive. *(Mixing Equipment, Inc.)*

drive in excess of rated capacity will ultimately lead to gear tooth surface distress and premature wear or breakage.

The gear reducer should be specifically designed for mixer service with the output shaft and bearings of sufficient size to carry the weight of the shaft and impellers. It must withstand all torsional, bending and thrust loads so that significant deflections are not transmitted to the gears and other bearings.

The thermal rating should not be less than the rated motor horsepower, and the mechanical horsepower rating should provide the service factor suggested in Table 7-6 as a minimum:

TABLE 7-6. AGMA Service Factors for Gear Reducers.

Service	Uniform Load	Medium Shock	Heavy Shock
Intermittent (2 hour/day)	0.90	1.00	1.25
Normal (10 hour/day)	1.00	1.25	1.50
Severe (24 hour/day)	1.25	1.50	1.75

All reducer bearings should be of the antifriction type and oil or grease lubricated. No bearings should be located below the drive base.

B-10 bearing life specifications (i.e., the number of hours of operation associated with a 10% probability of bearing failure, as determined by controlled mechanical testing under load conditions) should be related to the service life expected in the gears, as the purpose is to provide reasonable assurance of a given term of service before a major overhaul is required. An excessive B-10 specification usually forces the selection of an unnecessarily costly reducer. Typical B-10 bearing life exceeds 100,000 hours of operation. The relationship between bearing life and load is proportional to the increase in service factor raised to the 3.33 power. Thus if the Service Factor of a gear drive is increased by 30% life will increase by 240%.

Bearings positioned outside the reducer oil supply must be grease lubricated. Other bearings should be lubricated by a constant flow of oil through splash lubrication, an oil slinger or pump. Recommended lubricants should be in accordance with AGMA standards and available from local sources.

Helical and spiral-beveled gears supported by rolling element bearings operate with the highest efficiency of any major power transmission system. Power losses are usually less than 1.5% per reduction.

Each type of gear transmits power with an efficiency that usually decreases with greater speed reduction. Fig. 7-14 shows typical efficiencies for several types of gear drives. When multiple speed reductions are needed, the overall efficiency is the product of the individual efficiencies.

Worm gears are the most versatile gear drives in terms of speed variation, with possible reductions up to 100:1. However, the power-transmission efficiency of standard worm gearing is only 40–90%, depending on the amount of speed reduction. This is the lowest efficiency of the commonly used gear drives. Involute worm gears deliver slightly higher efficiency than standard designs.

Lost power is converted to heat and worm gears used for significant speed reductions need a high thermal rating to prevent overheating. Such gears are sensitive to oil level because either too much or too little can cause mechanical damage.

Planetary gearing has been often overlooked in the past because of its costs, but some newer planetary drives are very cost-competitive in agitation. These units have the highest efficiencies among the gear drives, and can offer more reduction than even worm gears—thus, applications that would require two or

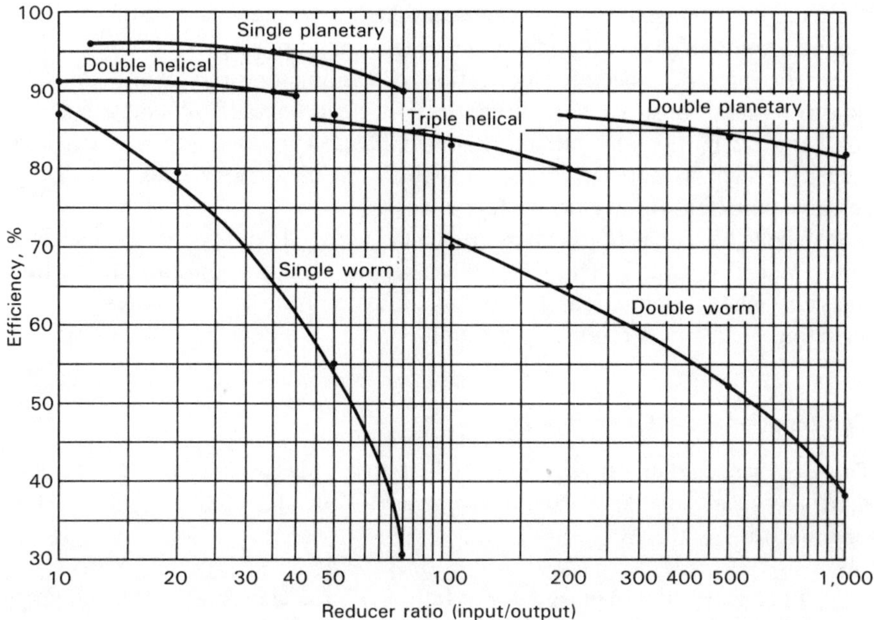

FIGURE 7-14. Typical speed reducer efficiencies.

more stages of other gears need only one planetary gear. Such drives should be considered for very low speeds, or in situations where mounting the motor coaxially with the impeller shaft has some special advantages (other drives require that the motor be offset).

Belt Drives

Because they are relatively inexpensive, belt drives are commonly used in applications where little speed reduction is needed; in side-entering agitators, for example. The conventional V-belt drive depends on tension in the belt, and face friction between the belt and sheave, to transmit power. Properly tightened, its efficiency is about 95%, but flexing and vibration loosen and deteriorate the belt.

The ultra-V belt is an improvement, in that it can carry higher loads with fewer belts and smaller sheaves, and costs less to operate. A more recent improvement is a toothed belt, or cog belt, whose teeth transmit power by meshing with the teeth of a cogged sheave-like a flexible gear. Mechanical efficiencies of such units are about 99%, and do not depend on proper tension. Thus maintenance cost is reduced.

Manufacturers claim a cog-belt life about four times that of V or ultra-V belts. As one would expect, such drives cost more than belts without teeth.

Since there is no metal-to-metal contact as the belt teeth mesh with the sprocket grooves, cog belt drives reportedly never require lubrication.

By eliminating the need for lubrication, contamination from oil drip, splatter, splash, or spray mist is also eliminated. The clean operating characteristics of cog belts are particularly attractive for application in contamination-sensitive industries such as paper mills. And there is no oil or grease to trap dirt or abrasive particles which can accelerate wear.

The belts have tensile members embedded with a flexible neoprene backing. These helically wound members provide tensile strength, prolonged flex life, and resistance to elongation. Takeup is eliminated from drive maintenance and operating precision is reportedly ensured.

Combination Belt–Gear Box Drive

This flex protected drive (assuming the belt drive is between gearbox and mixer shaft) provides many design and maintenance features for mixers. Some of these benefits are:

1. Flex protection. The agitator is fully flex protected; not only is the gearbox isolated from bending shock loads via the flexibly coupled, separately mounted bearing spool, but it is torsionally protected in addition.
2. Use of nonproprietary gearboxes. Agitators can utilize commercially available nonproprietary gearboxes. In many circumstances the customer can stipulate the gearbox manufacturer of their choice and, in fact, need not contact the mixer supplier for spare parts. This substantially reduces the cost of gearbox replacement and eliminates the problem of "marriage" to an agitator vendor for spare parts.
3. Ease of gearbox maintenance. Because of the modular concept of design, the gearbox can be maintained simply and with a minimum of downtime. By replacing the gearbox—an operation that involves only removing the belt(s), disconnecting the flexible coupling, and unfastening four bolts—all bearing and gear maintenance can be accomplished quickly by unskilled maintenance personnel. The agitator shaft and turbines are never touched.
4. Ease of bearing maintenance. The separately mounted bearing spool is a discrete cartridge containing the upper agitator shaft and shaft bearings. This cartridge can be removed quickly and easily *without removing the agitator from the tank or disturbing the gearbox* with the use of the shaft holding tool. By having a spare cartridge available, the agitator shaft bearings can be completely changed with no setup or skilled labor requirement. In addition, the old bearings can be retrofitted with a new set in the comparable comfort of the maintenance area.

5. Process flexibility. By using a belted secondary reduction stage, agitator speed adjustments associated with process changes can be easily, quickly, and inexpensively handled by changing a belt sheave. Minimum service factors of 1.50 (AGMA) on the gearbox and 1.30 (RMA) on the belt drive allow for HP increases of 30% without sacrificing mechanical integrity and with only a sheave and motor change.
6. Rugged design. To minimize maintenance, the number of moving parts of the agitator has been kept to the minimum. The housing and straight through shaft are designed to accommodate higher power inputs, if required. Heavy duty "off-the-shelf" bearings with a B-10, 100,000 hour life expectancy, are standard in all bearing locations including the gearbox.
7. Modular design. Normally the shaft length dictates the diameter and housing size rather than the torque and rotational speed. With modular design one can independently tailor the housing to carry a suitable shaft and the drive to provide the necessary power input. As a result, a grossly oversized gearbox is not required to carry an integral large diameter output shaft, as is the case with totally gear driven machines. The flexibility allows us to precisely design a machine to a requirement of shaft size and power input at a very attractive price.

Required Tension in V-Belts

Running under load causes that section of V-belt under the load to be tight, under tension T_1, while the other side is slack under tension T_2. The difference between these two tensions, $(T_1 - T_2)$ is the effective tension, T_e, which is required to turn the driven sheave.

From this formula, it is evident that $T_1 = T_e + T_2$, and T_1 must be greater than T_e by an amount not only sufficient to wedge the V-belt into its sheaves but also to overcome the effect of centrifugal force that tends to lift the belt away from its sheaves.

If the difference between the tight-side and slack-side tensions is expressed as a ratio R:

$$R = T_1/T_2 \tag{37}$$

It is possible to relate T_1 to T_e by the expression:

$$T_1 = T_e\,(R/(R - 1)) \tag{38}$$

Furthermore, T_e can be expressed in terms of the power requirements as:

$$T_e = 33{,}000\,(\text{Hp})/S \tag{39}$$

$$S = (0.262)\ (\text{PD})(N) \tag{40}$$

where: Hp = the total design horsepower per belt
S = belt speed, ft/min
N = sheave speed, rpm
PD = the sheave's pitch diameter, in.

From these, it is apparent that T_1 can be calculated in terms of horsepower, rotational speed, and sheave pitch diameter, providing there is a value for the tension ratio R.

Ideal tension ratios based on the arc of contact of the sheaves, have been established through experiments. These have been incorporated, with the other variables into a nomograph (see Fig. 7-15) which can be used not only to calculate the belt tension, but also the diameter of a companion sheave when the

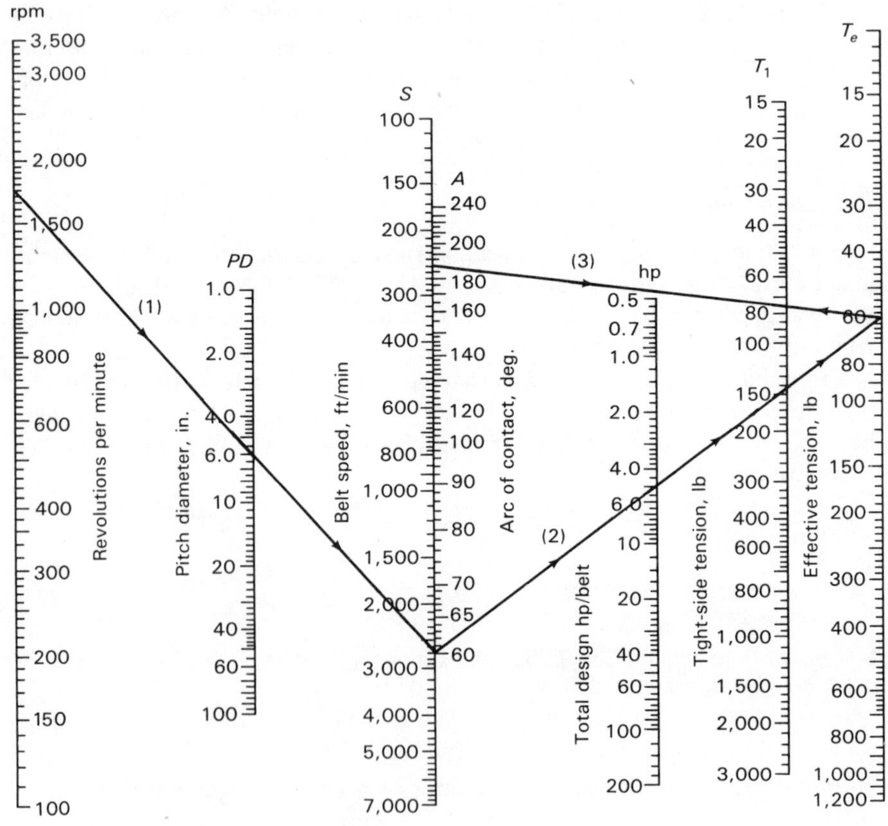

FIGURE 7-15. Nomograph to determine belt tension for V-belt drives.

diameter of the other sheave and the rotational speeds of the driver and driven machine are known.

A measure of percentage elongation probably offers the simplest means of setting and adjusting the tension of single V-belts. With the belt in place and slack removed, a steel tape is wrapped around the outside of the belt and the length recorded. Then an elongation factor (ranging from 0.01 for new A, B, and 3V belts and 0.0075 for new C, D, E, 5V and 8V belts to 0.005 for used A, B and 3V belts and 0.0025 for used C, D, E, 5V and 8V belts) is applied to the measured length to establish the length under tension. This method is not to be used with glass cord or steel cable belts.

The relation of elongation factor to pounds tension, as well as other methods for measuring tension, can usually be found in the belt-manufacturer's catalogue.

Example k. A 1,750 rpm motor delivers 5 hp via a 6 in. diameter sheave and a V-belt to a machine that also has a 6 in. sheave. What is the required belt speed? the effective belt tension? the tight-belt tension? and the slack-belt tension? The arc of contact (A) on each sheave must equal 180 degrees, since they are of equal diameters. On the nomograph, align 1,750 rpm with 6 in. on the pitch diameter scale, extend the line to the S scale and read the speed as 2,770 ft/min. Align this speed with 5 on the Hp scale, extend the line to the T_e, scale and read 60 lb. Align $T_e = 60$ with 180° on the A scale and read $T_1 = 75$ lb; $T_2 = (T_1 - T_e) = 15$ lb; and the total load on the shaft is $(T_1 + T_2) = 90$ lb.

Where multiple belts are used, divide the total load by the number of belts.

F. SHAFT SEALS

Mechanical Seals

A mechanical seal will only work properly and dependably if it is a part of a well designed drive. A double mechanical seal is illustrated in Fig. 7-16. A single seal arrangement is the bottom half of the double seal. The detail of the rotating and stationary seats of the seal are shown in Fig. 7-17. The rotary element is pinned to the shaft while the stationary element is held in place by the mounting flange and O-ring as shown.

A critical concern when it comes to design and operation of a mechanical seal is deflection of the shaft in the seal area. Calculated deflection for a given bearing spacing must be considered. Fig. 7-18 is a schematic of a mixer shaft with a mechanical seal. As shown, the seal is located x inches below the inboard shaft bearing. A force F acts at the end of an overhung shaft.

The bearing load may be calculated by:

$$\text{LOAD} = (FL/A) + F \tag{41}$$

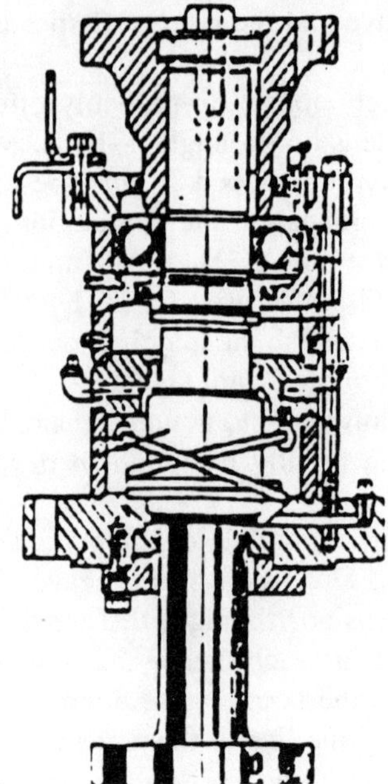

FIGURE 7-16. Double mechanical seal. *(Mixing Equipment, Inc.)*

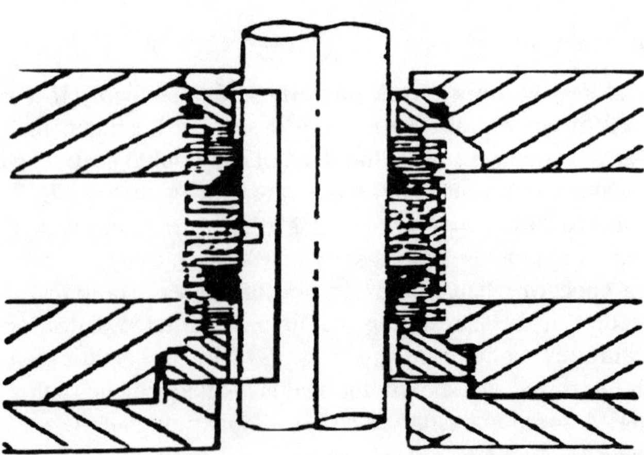

FIGURE 7-17. Single mechanical seal. *(Mixing Equipment, Inc.)*

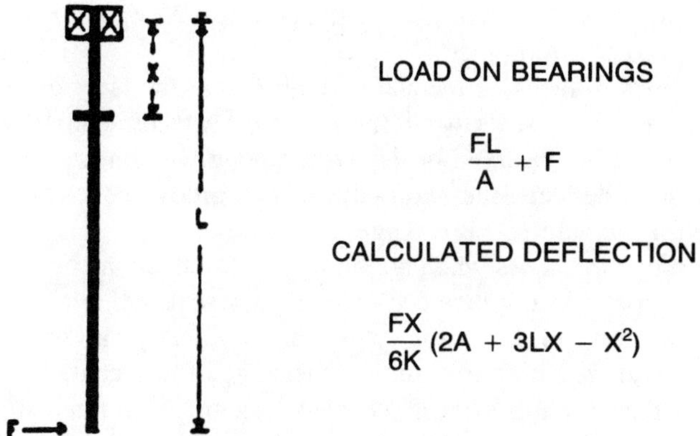

FIGURE 7-18. Mixer shaft with mechanical seal. *(Mixing Equipment, Inc.)*

where: F = hydraulic force acting on the impeller, lbs force
L = shaft length from inboard bearing to impeller, inch
A = bearing spacing, inches

Deflection in the seal area may be calculated as follows:

$$\Delta = \frac{FX}{6K}(2A + 3LX - X^2) \tag{42}$$

where: K = constant
Δ = Shaft deflection, inches
X = Distance from inboard bearing to seal inches

The radial load on the bearings is greater for shorter bearing spacings A. For the same forces a shorter bearing span will require larger bearings. Conversely, the same bearing will have a shorter life.

A shorter bearing spacing will result in a smaller calculated deflection and normally less seal wear. Seal life is not predicated on calculated deflection per se, but on how the calculated deflection compares with the allowable deflection of the seal faces based on mechanical seal design limits.

Seal Faces

''Face-on-face'' sealing occurs when the rotary and stationary elements are the same width in the area of common contact. This concept is in stark contrast to the mating of a narrower rotary face with a wider stationary face. Off-the-shelf pump seals are similar in design to the latter concept. An agitator seal, however, must be designed "face on face."

The problem with faces of unequal width is that as the faces of the stationary and rotary elements or seats wear (usually rotary faces tend to wear faster, although wear can be observed on both), a groove will form on the stationary face. As the shaft deflects, the outer edge of this groove gets rounded off, creating a path for premature seal leakage.

A severe problem occurs when abnormal shaft deflections occur. Given this situation, the groove in the face of the stationary seat will attempt to hold the rotary face, resulting in the brittle rotary face (typically carbon) getting fractured at the corner, resulting in significant leakage of lubricant. Lubricated mechanical seals have the lubricant in a closed system, pressurized slightly above the maximum operating pressure in the tank. This is to ensure that the seal faces are in compression with one another under normal operating conditions. When tank pressure exceeds lubricant pressure or when seal faces wear or break, lubricant will leak from failure of the seal.

With a narrow faced mechanical seal (where the width of the rotating and stationary faces are not the same), shock loads requiring a high service factor speed reducer and causing significant shaft deflection in the seal area will likely fracture the rotary seal face after wear has occurred.

Package versus Cartridge Seals

Deflection in the seal area will affect seal life. Actual deflection may be higher than that calculated due to machining tolerances of components, particularly in units having short bearing spacing. Many short bearing span gearboxes are used in conjunction with a mechanical seal supplied on a sleeve. O-rings seal the sleeve to the mixer shaft. The problem with "seal on a sleeve" designs is that misalignment can occur between shaft and bearing centerlines due to accumulated tolerances between the bearing on the sleeve, which in turn is on the shaft sleeve, which is turn is on the mixer shaft, with an additional gap required for the o-ring. The "seal on a sleeve" design is often called a seal package. A "built-in deflection" is inherent in this design due to manufacturing tolerances of the components in the package (i.e., bearings, sleeve, mixer shaft, etc). This in concert with the actual shaft deflection caused by hydraulic fluid force can prematurely fail the mechanical seal.

Gearbox reducers with larger bearing spacings and those utilizing a tubular shaft design (i.e., hollow quill concept) eliminate the need for a "seal on a sleeve" package. Seal deflection is minimized in this type of drive because:

1. True alignment between bearing and mixer shaft is ensured because a fixed bearing above the rotating seats is pressed onto the seal shaft.
2. Tolerance buildup associated with sleeve seals is nonexistent.

This seal assembly is referred to as a cartridge design, since the entire assembly can be pretested at operating speed and temperature before mounting it on the tank nozzle.

The cartridge contains the mechanical seal itself, bearing, and shaft with mounting flanges on both ends. The mechanical seal is preinstalled on the shaft. Downtime is minimized when installing the cartridge since it bolts as an entire assembly to an upper flange, lower flange and to the tank nozzle flange. Fig. 7-19 illustrates the mechanical seal cartridge assembly.

The "seal on a sleeve" package design is lighter in weight than a cartridge because it does not have a shaft and coupling half. In most cases however, this advantage is far outweighed by the misalignment disadvantage.

Removal of a cartridge seal assembly is accomplished by lifting the assembly and lower mixer shaft up until an intank coupling clears the mixer mounting nozzle. The entire unit must be held in place with some sort of shaft holding device. Once the lifted seal/mixer shaft assembly is secured, the cartridge may be disconnected at its flanges for removal and installation of a new cartridge. With the new cartridge in place, the seal/shaft assembly can be lowered and the seal unit can be bolted to the mixer mounting flange. By replacing the entire cartridge housing, with its preinstalled mechanical seal, alignment is ensured

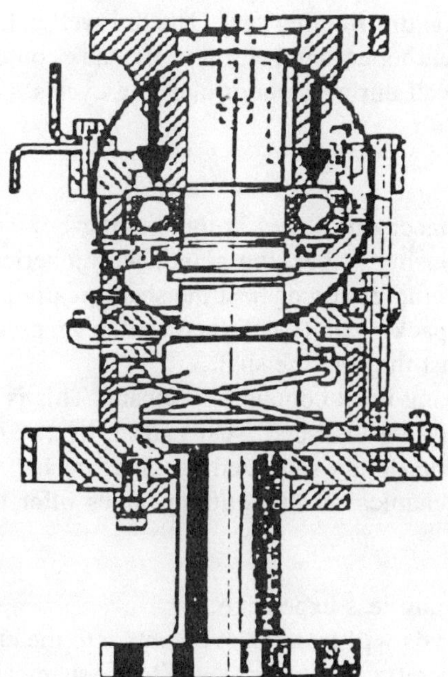

FIGURE 7-19. Mechanical seal cartridge assembly. *(Mixing Equipment, Inc.)*

and downtime minimized. The faulty mechanical seal can be repaired in the shop, while the newly installed unit is in operation. A step-by-step approach to changing a seal cartridge is shown in Fig. 7-20. The particular drive assembly happens to be a belt drive/gearbox unit.

Dry Running Mechanical Sales

For those manufacturing processes that must be conducted in orderly, precisely planned sequences with critical time parameters, or if all products must be kept completely free from contamination from any source, a seal without lubricant is preferred. Dry running seals eliminate the potential of lubricant leaking into a process vessel in the event of seal failure. Downtime and product loss is minimized.

Dry running seals are ideal for maintaining sterile operating conditions, such as those required in the food, pharmaceutical, and biotechnology industries. They can be installed as single or double mechanical seals, just like their lubricated counterparts.

The dry running seals can be installed in a cartridge assembly, ready for retrofitting lubricated seal cartridge units.

To eliminate possible infiltration of microscopic particles of material produced by seal face contact with the product being mixed, an optional dry well can be provided in the dry running seal. The dry well collects and isolates face wear particles, preventing contamination of the tank contents. Pipe taps permit flushing of the dry well during agitator cleaning cycles.

Stuffing Box Seals

An alternative to a mechanical seal is the stuffing box seal. The stuffing box seal consists of a housing or packing gland, with a series of rings of packing around the shaft in compression against the shaft creating a seal. A lantern ring fits over the rings of packing and can be screwed down pressing the rings against each other and against the rotating shaft.

The rings of packing must contain a lubricant. This is to prevent erosion of the ring due to excessive heat and wear caused by the rubbing action of the mixer shaft against the tightly squeezed packing rings.

Compared to mechanical seals, stuffing boxes offer these advantages and disadvantages:

1. Stuffing boxes are less expensive.
2. Stuffing box seals will leak tank contents into the environment (even under normal operating situations), while mechanical seals will not leak. Leakage is minimal under normal operating conditions and with well maintained stuffing boxes. Leakage rates will increase under abnormal

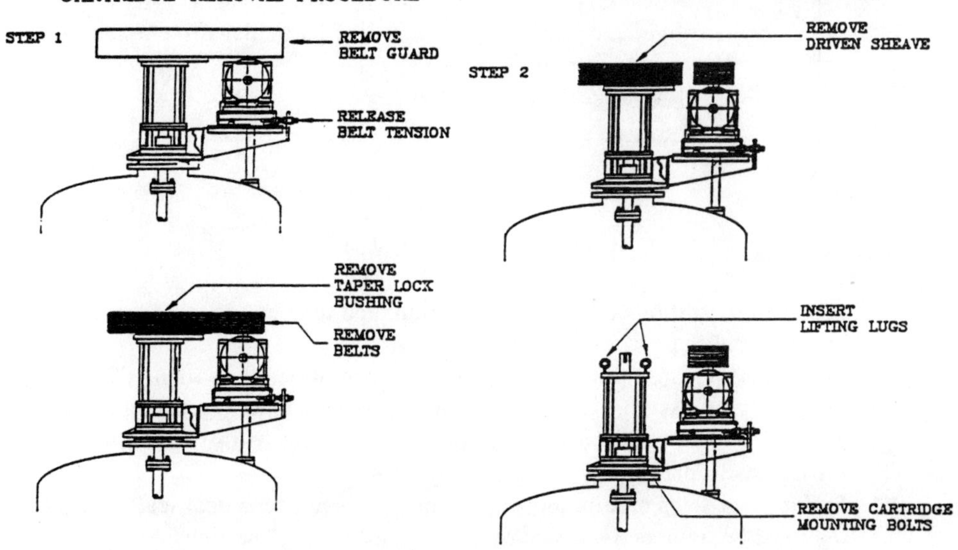

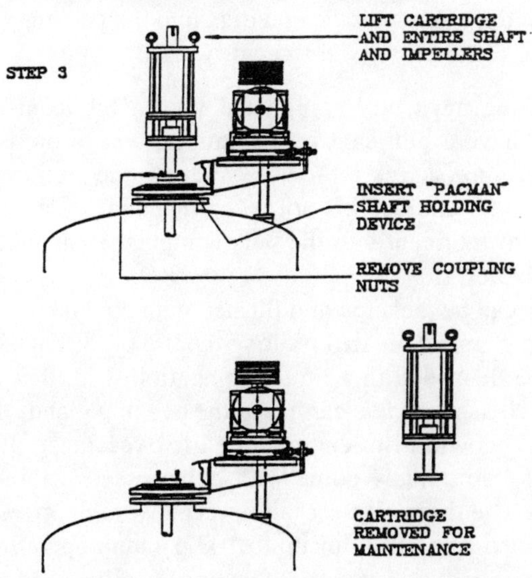

FIGURE 7-20. Cartridge seal removal. *(Prochem Mixing Equipment, Inc.)*

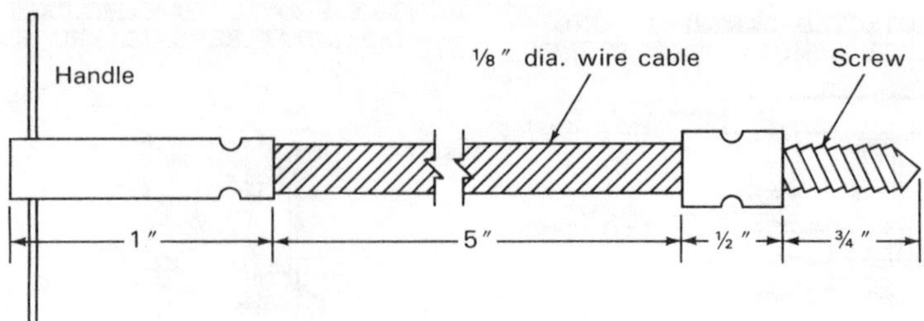

FIGURE 7-21. Packing ring puller.

operating conditions of increased pressure and temperature or if the stuffing box seal is poorly maintained.

3. Stuffing boxes are limited to low pressure operations (less than 100 psi). Under high pressure, leakage becomes uncontrollable. Mechanical seals can be installed up to 10,000 psi operating pressure, with special staging provisions to pressurize the seal faces.
4. Stuffing box seals require more maintenance than mechanical seals. Packing must be tightened daily during mixer operation. The rings are subject to wear faster than seal faces in a mechanical seal and therefore must be replaced more frequently. Replacing worn out packing in a stuffing box is difficult because the clearance between shaft and packing is small. Fig. 7-21 shows a tool that is very useful in removing packing from stuffing boxes. The wire cable gives the mechanic flexibility in getting at the packing, and the screw at the end digs into the packing to help pull it out. Such a tool can be easily fabricated.

Rings of packing must be impregnated with a lubricant such as teflon or graphite or must have a lubricant in the stuffing box to protect the rings from wear. The former approach is preferable, making the stuffing box a "dry running seal." The latter approach is not recommended, because it will assuredly promote leakage of lubricant into the tank contents. Remember, stuffing boxes inherently leak by design.

Stuffing boxes can be jacketed and flushed with coolant to keep temperatures down in the sealing area. This will prolong the life of the packing rings. Stuffing boxes can also be flushed with a liquid compatible with that in the tank to act as a coolant as well as provide cleaning of the packing gland, lantern ring, rings of packing, etc. This will also lengthen the life expectancy of packing rings.

Stuffing boxes traditionally come with either a set of three or set of seven rings of packing. The three ring set is for very low tank operating pressures up to 25 psi. The seven ring set is for up to 100 psi tank operating pressure. Both designs can be supplied with optional flushing, cooling, and lubricating provisions previously discussed.

Glossary of Terms

Aerobic Fermentation processes requiring moderate to high oxygen mass transfer

Agglomerate To form into a ball

Agitator A machine assembly for mixing, i.e., mixer, stirrer

Airfoil A curved surface impeller with blades similar to an airplane wing in cross section to maximize pumping and minimize shear. See Fluidfoil and Hydrofoil.

Ammeter Instrument to measure amperes drawn by a motor

Anaerobic Fermentation processes requiring low oxygen mass transfer rate

Anchor Radial flow impeller of a design resembling a horseshoe, running close clearance to the inside tank wall for moderate to high viscosity applications

Angle of Attack Angle formed by the resultant axial and radial velocity vectors with the trailing edge of the impeller blade.

Apparent Viscosity The viscosity of fluid being mixed within the periphery of the impeller, at the shear rate of the impeller.

Axial Flow Fluid movement in the direction of the axis of rotation of the mixer

Baffles Vertical flat plates installed on the inside tank walls to transform tangential swirling flow patterns into vertical, non-swirling flow.

Bar Turbine A high-shear, low-flow radial disk turbine with rectangular pieces of bar stock, alternating top and bottom on the disk

Bending Moment Product of force or hydraulic load (i.e., fluid force) and shaft length to determine loads acting on mixer components along its shaft (i.e., mounting flange, bearings, etc.)

Bioreactor Same as fermenter

Blade Passing Frequency A shaft excitation force equal to the number of impeller blades times operating speed

Blend Time Mixing time required to achieve desired final product blend consistency.

Blending The merging or mixing of components into a final product of different properties

Brookfield Viscosimeter An instrument to measure viscosity and shear utilizing spindles running at various controlled speeds
Broth Slurry in a fermenter

Camber Radius of blade curvature on a hydrofoil
Chevron Impeller Radial flow turbine with V-shaped blades for cone bottom tank mixing
Coalesce Smaller gas bubbles forming into larger ones
Cog Belt A ribbed or toothed belt used in belt drives to transmit power without slippage
Conduction Heat transfer through walls of tubes, between wall of tank and jacket, etc.
Convection Dynamic heat transfer created by turbulent fluid velocity across heat transfer surfaces
Critical Speed The speed of a rotating shaft at which the forces acting on it give rise to large vibrations producing natural harmonic frequencies of bending in the shaft

Draft Tube A pipe or tube suspended in a tank, submerged beneath the liquid level, incorporating a mixing impeller within it.
Drag Coefficient A constant reflecting the resistance to movement brought to bear upon an impeller by the fluid through which it rotates
Drive Train The speed reduction component of a mixer i.e. gear box, belt drive
Dynamic Similarity All force ratios are constant on scale-up
Dynamometer An instrument for measuring torque transmission in a mixer

Emulsion A dispersion of droplets or globules of one liquid into another immiscible liquid
Entrained Flow Flow incorporated from its surroundings by a mixing impeller
Equivalent Weight For shaft critical speed calculations, it is the weight equivalency of all impellers taken into consideration as a single turbine

Fermenter A vessel for growing microbial cells and biological products using oxygen and nutrients for growth
Fillet An accumulation of settled solids in the corners of mixing tanks
First Natural Frequency The shaft speed at which damaging harmonic vibrations occur
Flow Volumetric flow rate developed by a mixer i.e., cubic feet per minute or gallons per minute
Flow Induced Loads Dynamic forces imparted on baffles by rotating impellers
Fluidfoil Same as hydrofoil impeller; see airfoil
Fluid Mechanics Flow and shear development of a mixing impeller
Fluid Forces Hydraulic load imparted on the impeller blades by the tank surroundings
Fluid Stress The work performed by forces created by fluid shear
Free Settling Solids Particles with terminal settling velocities greater than 1 ft/minute
Frictional Loss Head loss in a draft tube circulator due to frictional drag of the fluid against the tube wall

Gas Dispersion A mixture of gas and liquid

Gates Radial flow impeller with multiple vertical blades mounted on two horizontal blades

Geometric Similarity For scale-up, all pertinent dimensions are similar and have constant ratios

Geysering Burping flow pattern observed in a gas dispersion mixing application when isothermal gas power exceeds mixing power

Head Fluid pressure applied or developed by a pump or agitator; synonymous with shear rate developed for a mixer

Heat Transfer Exchange of energy between fluids of different temperature

Hindered Settling Solids Particles with terminal settling velocities less than 1 ft/minute

Homogenizers/Colloid Mills High-shear, low-flow dispersion mixers for uniform dispersions of fine solids in liquids

Hydraulic Optimization Choosing the most cost effective combination of mixing torque, turbine type, diameter and speed for flow controlled applications

Hydrofoil An airfoil impeller used in liquid mixing applications, see fluidfoil and airfoil

Impeller A bladed assembly mounted on a shaft to provide the driving force for mixing, i.e., turbine, propeller

Inertial Load Force imparted on a mixer with its impeller imbedded in settled solids when the motor is turned on

Intimate Gas Dispersion Mixer power a multiple of 1–3 times the isothermal gas power creating small gas bubbles evenly distributed across the liquid surface

Isothermal Gas Horsepower Work associated with the expansion of gas introduced into a stirred tank for dispersion

Kiel Probe A device to measure flow and head loss in draft tube circulators

Kinematic Similarity All velocities have a common, constant ratio on scale-up

Laminar Flow The regular, continuous smooth movement of fluid with Reynolds number less than 100

Laser Doppler Anemometry Measuring velocity of a particle in a fluid with scattered laser light beams

Laser Velocimeter Particle velocity measurement by Laser Doppler Anemometry

Launder An open trough

Leaching Extraction of a soluble substance from a solid by the action of a liquid solute moving through the pores of the solid

Macroscale Pertaining to bulk fluid motion

Mass Transfer Exchange of mass between fluids of different concentration and properties

Mechanical Seal An alternative shaft sealing device to the stuffing box, utilizing spring-loaded sealing faces (rotary and stationary) to seal the tank contents from escaping along the mixer shaft

Microscale Turbulent mixing in a finite small incremental volume of fluid

Minimum Gas Dispersion Mixer power barely compensates for isothermal gas power while distributing gas across the liquid surface
Momentum The product of the mass of a particle and its velocity
Momentum Flux The product of mixer flow rate and fluid velocity

Non-Newtonian Fluids exhibiting a changing viscosity, as a function of shear rate
Normal Perpendicular

Orifice A round plate with a sharp edged, beveled hole inserted in a pipe to determine pressure drop
Ott Meters A velocity measuring device incorporating a small shaft-mounted propeller that spins in a fluid in motion

Pachuca An airlift draft tube circulator mixer
Paddles Flat bladed radial flow impellers, usually with two or more blades angled at 90° to the horizon
Performance Curve Plot of head versus flow development of a draft tube circulator
Phase Inversion The phenomenon of having the lighter and heavier phases of a dispersion switch roles after settling and separation, i.e., the lighter phase is on the bottom and heavy phase on top—due to excessive mixer shear during emulsification
Pilot Plant Small scale simulation of a larger scale mixing system
Pitch Angle that a mixing impeller blade makes with the horizon
Pitot Tube An instrument to measure velocity (and inherent velocity head loss) in a draft tube circulator
Platecoil Vertical baffles with tubes embossed in them for heat transfer applications—also known as panel coils
Plug Flow Turbulent pipe flow without longitudinal or axial mixing
Power Number Dimensionless value as a function of impeller geometry, type, number of blades and Reynolds number
Process Result The design premise of the mixer stated in process terms or requirements
Process Torque A measure of work done on the fluid being mixed; it is delivered power divided by rotational speed
Propeller An axial flow impeller used primarily on side-entry and fractional horsepower portable and fixed mount mixers
Proplets Vertical blade tips on the extreme outer edge of hydrofoil impellers to eliminate flow recirculation at the blade tip, and enhance overall flow efficiency
Pseudoplastic Viscosity relationship indirectly proportional with shear rate
Pump Casing Impeller housing for a pump
Pumping Number Relative index and measure of pumping efficiency

Radial In the direction of the radius of a mixing impeller i.e., perpendicular to the axis of rotation
Rakes Radial flow impeller with multiple vertical blades off one horizontal blade
Reactor Kinetics Dynamics of substances undergoing chemical change

Resuspension Vents Vertical slots in a draft tube to promote resuspension of settled solids

Retrofitting Replacing a less efficient mixer impeller with a more efficient one to improve process results

Reynolds Number Dimensionless group associated with a flow regime being laminar or turbulent

Riser A dip tube outlet for the discharge of a continuous flow of slurry from one tank to the next in a series of tanks

Rushton Turbine Radial flow flat-bladed disk turbine

Saw Tooth Impellers A disk turbine with sharp teeth on the periphery of the blade used to shear particles being mixed

Scale-up Taking mixing results and parameters observed in a smaller scale into a larger scale, while keeping results constant

Shaft Shear Stress The result of dynamic forces acting on a mixer shaft and is the resultant of torsional and bending stresses (i.e., the square root of the sum of the squares of torsional and bending stress)

Shear Fluid layers in contact with each other, sliding across each other parallel to the plane of contact

Solidity Ratio Ratio of area of impeller blades to circular area of the impeller diameter

Solids Attritioning Breaking down of solids into finer particle size

Solids Suspension A mixture of solids in liquid(s), i.e., a slurry or dispersion

Solvent Extraction The selective separation of one or more components of a mixture by the use of a solvent liquid

Sparger A device for introducing gas into an agitated vessel, usually below the impeller

Specific Gravity Dimensionless relationships of the ratio of the density of a liquid to that of water

Spiral Axial flow impeller resembling an auger with one or two helical flights for moderate to high viscosity applications

Stall Point Discontinuity region on draft tube circulator performance curve where power is wasted in conversion to turbulence

Static Inline Mixer A pipeline mixer without moving parts

Stator Vanes Flow straightening partitions downstream of the draft tube circulator impeller to streamline flow and minimize head loss

Steady Bearing A bushing at the end of a mixer shaft having the shaft fit tightly into it, as if the bushing were a bearing to steady the shaft from vibrations

Stratified Blending Bringing to a uniform blend different strata or batches

Stuffing Box A shaft sealing device, flange mounted to the tank, incorporating compressed packing to seal the tank contents from escaping along the mixer shaft

Superficial Gas Velocity Volumetric flow rate of gas divided by cross-sectional area of the mixing vessel

Thermal Effectiveness Factor A constant reflecting heat transfer based on surface geometry and how well fluid in turbulent motion moves over and around the particular surface

Thrust The reactive force opposite to the direction of flow; proportional to fluid momentum

Torque The force that acts to produce rotation

Torque Cell A device for measuring torque transmitted along a mixer shaft, and hence power

Transverse Uniformity Complete uniformity throughout an entire cross section of a static pipeline mixer

Turbine See impeller

Ultimate Weight % Settled Solids Percent by weight of settled solids in a tank with the supernatant liquid above the bed of solids drawn off

Uniform Gas Dispersion Mixer power, a multiple greater than three times that of the isothermal gas power, creating very small gas bubbles, dispersed across the surface of the liquid in a uniform manner

V-Belt A type of multiple belt used in belt drives for a mixer with a cross sectional shape of a "V"

Viscosimeters Instruments to measure viscosity and shear rate

Viscosity A measure of fluid "thickness"; the quotient of shear stress and shear rate

Vortexing Deep swirling, circular flow pattern with eddy current formation created by mixing without baffles

Wattmeter Instrument to measure power drawn by a motor

Wiper Attachments Extensions to the blades of anchor impellers to scrape the wall of the vessel during rotation; usually hinged to the blades and often made of Teflon

Bibliography

Backer, Mark P., Metzger, Linda S., and Kao, Eugene I. Shear and Mixing in Large Scale Suspension Culture presented at 80th Annual Meeting of AIChE, New York, Nov. 15–20, 1987.

Bjurstrom, Edward E., Jost, John and Swartz, James R. 1985. Operational Parameters and Broth Properties Affecting Oxygen Transfer in Bioreactors. Presented at Biotech '85 in USA.

Brain, T. J. S., and Man, K. L. Heat Tranfer in Stirred Tank Bioreactors. Presented at AIChE Annual Meting, New York, Nov. 15–20, 1987.

Bryant, J. The Characterization of Mixing in Fermentation. *Adv. in Bioeng.* **5,** ed. Ghose, et al., pp. 101–123. New York: Springer-Verlag, 1975.

Calabrese, Richard V., and Stouts, Carl M. 1989. Flow in the Impeller Region of a Stirred Tank. *Chemical Engineering Progress*. May 1989: 43–50.

Caplan, F. Specific Gravities of Slurries or Mixtures. In *Calculation and Shortcut Deskbook*, pp. 37. New York: McGraw-Hill, Inc.

EMI Incorporated. Article TP-101 Parameters for Mixing Equipment Specifications. Article TP-102 Mixing Process Classifications. Article TP-103 Basic Design Calculations. Article TP-104 Mixer Impellers. *Agitation Design Guidelines and Fundamentals*: 1–14.

EMI Incorporated. Article CSM-2. Sizing the Cleveland Static Mixer: 1–5.

Garrison, Charles M. 1981. How to Cut Agitation Costs. *Chemical Engineering* Nov. 30, 1981: 73–78.

Gbewonyo, K., DiMasi, D., and Buckland, B. C. Characterization of Oxygen Transfer and Power Absorption of Hydrofoil Impellers in Viscous Mycelial Fermantations. In *Biotechnology Processes Scaleup and Mixing*, pp. 128–134. New York: American Institute of Chemical Engineers, 1987.

Godfrey, J. C., Hanson, Carl, Slater, M. J., and Tharmalingam, Shanmagam. Studies of Entrainment in Mixer Settlers. *American Institute of Chemical Engineers Symposium Series 173,* **74:** 127–133, 1978.

Ho, Chester S., and Oldshue, James Y. *Biotechnology Processes Scaleup and Mixing*. New York: American Institute of Chemical Engineers, 1987.

Knaebel, Kent S. Simplified Sparger Design. In *Calculation and Shortcut Deskbook*, pp. 104. New York: McGraw-HIll, Inc.

Lally, Kenneth S. A-315 Axial Flow Impeller for Gas Dispersion. In Technical Report #A315-2 Mixing Equipment Co., Inc. Rochester, NY, 1987.

LIGHTNIN Technology Seminar. Section 2A Mixing Theory and Practice: Blending. Section 2B Solids Suspension. Section 2D Heat Transfer. Section 2E Gas-Liquid Operations. Section 3 Standard Mixer Selections. Section 4 Mechanical Design. Section 5 Scale-up and Pilot Planting.

McClarey, Michael J., and Ali Manssori, G. Factors Affecting the Phase Inversion of Dispersed Immiscible Liquid-Liquid Mixtures. *American Institute of Chemical Engineers Symposium Series 173*, **74:** 134–139, 1978.

McDonough, Robert 1986. Use of Energy-Efficient Hydrofoil Impellers in Chemical and Food Processing. Presented at Chemical and Food Processing Table Top Show, May 14, 1986, Niagara Falls, NY.

Michel, B. and Miller, D. Power Requirements of Gas-Liquid Agitated Systems. *A.I.Ch.E. Journal*, 1962.

Miller, D. Scaleup of Agitated Vessels, Gas-Liquid Mass Transfer. *AIChE Journal*, **20:** 445, 1974.

Nienow, Alvin W. Gas Dispersion Performance in Fermenter Operation. *Chemical Engineering Progress*, February 1990: 61–71.

Olderstein, A. J. Comparison of LIGHTNIN A310 Laser foil Impeller to Prochem Maxflo T Impeller. In Technical Report #A310-6, Mixing Equipment Co., Rochester, NY, 1984.

Oldshue, James Y. Current Situation in Fluid Mixing for Fermentation Processes. In *Biotechnology Processes Scaleup and Mixing*, pp. 3–5. New York: American Institute of Chemical Engineers, 1987.

Oldshue, James Y. Fluid Mixing in 1989. *Chemical Engineering Progress*, May 1989: 33–42.

Oldshue, James Y. Fluid Mixing Technology and Practices. *Chemical Engineering*. June 13, 1983: 82–108.

Oldshue, James Y. New Concepts in Slurry Storage.

Ruiz, Luis Rios. Dry Running Seal Retrofit Prevents Contamination of Products. *Pharmaceutical Engineering* **10**(1): 34–36, 1990.

Rushton, J. H., and Oldshue, J. Y. *Chemical Engineering Progress*, **55** (25): 181–198, 1959.

Salzman, Ronald N., Webster, Walter C., and Lally, Kenneth S. Structural Composite Mixers. *Chemical Engineering Progress*. November 1988: 50–55.

Schwartzberg, Henry G. and Treybal, Robert E. Fluid and Particle Motion in Turbulent Stirred Tanks. *Industrial and Engineering Chemistry Fundamentals* **7**(1): 1–6, 1968.

Scott, Roy R. Emulsions. *Practical Aspects of Emulsion Technology*, October 21, 1987.

Sekar, Chandra. Tool Removes Old Packing. In *Calculation and Shortcut Deskbook*, p. 66. New York: McGraw-Hill, Inc.

Shaw, John A., and McDonough, Robert J. Mechanical Agitation in the Carbon-In-Pulp

Process. Presented at International Mining Conference, Perth, Australia, June 1, 1982.

Timmermann, D. N. Streetsmarts: How to Specify and Select Gear Drives. *Chemical Engineering Progress*, March, 1989: 15–18.

Tunison, M. E. and Chapman, Thomas W. Characterization of the Kenics Mixer as a Liquid-Liquid Extractive Reactor. *American Institute of Chemical Engineers Symposium Series 173,* **74:** 112–119, 1978.

Tyler, J., Henry, W., and Park, J. Characterization of Hydrofoil Impellers in Large-Scale Fermentation Processes.

Uhl, Vincent W., and Von Essen, John A. Scaleup of Fluid Mixing Equipment. In *Biotechnology Processes Scaleup and Mixing*, pp. 155–167. New York: American Institute of Chemical Engineers, 1987.

Uhl, Vincent W. and Gray, Joseph. *Mixing Theory and Practice*. Orlando, Florida: Academic Press Inc., 1986.

Index